CREATION, EVOLUTION, REINCARNATION, AND OUR DESTINY

DESTINY, Volume Five

by EDWARD NAGELL

ISBN 0-7414-4752-5

Published by:

INFINITY
PUBLISHING.COM

1094 New DeHaven Street, Suite 100
West Conshohocken, PA 19428-2713
Info@buybooksontheweb.com
www.buybooksontheweb.com
Toll-free (877) BUY BOOK
Local Phone (610) 941-9999
Fax (610) 941-9959

Printed in the United States of America

Printed on Recycled Paper

Published May 2009

Dedicated to my darling wife, my editor, advisor,
friend, and sweetheart.

And to all the preachers, friends, authors, and critics
who helped me to live a satisfying life.

TABLE OF CONTENTS

INTRODUCTION

Religion and Science

Many, probably most, scientists think that their beliefs directly oppose religious beliefs. On the other hand, many religious people think that scientists are destined to go to hell. Neither attitude is constructive. Science and religion should be partners in bringing humanity out of darkness, not criticizing one another's efforts to do that. If science ever solves all the technical problems and religion ever gets complete insight into what creation is all about, both scientists and religionists will be talking about the same thing. At one time, philosophers believed they were the same.

Many centuries ago, science *was* part of religious activities. Priests took care of medical problems. Religious schools trained scientists and engineers. Gradually, science and engineering split away from religious leaders; but the terms covered all the fields of investigation that split into many specialties. Religion has also gradually split apart into thousands of different philosophies. We should consider science in general and religion in general as two areas that split apart just as physics and chemistry separated into two fields that have split into separate specialties.

Science and engineering advancements have made life a lot easier, safer, and more productive. Most people admit that, regardless of their religious beliefs, except for those that cling to remnants of some previous time that they believe was better. Scientists have been reluctant to admit that most religions have also made life a lot easier, safer, and more productive for those that followed the rules of their religions. Conflicts have risen when leaders of either side thought that a new idea or concept threatened their authority. There are many examples in technical fields where authorities rejected new ideas that they thought would make the authorities obsolete. Similar attitudes have kept religious leaders from accepting new theories and philosophies about religious things. Prejudice, pride, greed, and selfishness in leaders from both areas continue to hamper advancement of knowledge and understanding about the destiny of human beings.

One problem with the concern for our destiny is many, perhaps most, scientists seem to think that we have no destiny past the grave. They do not think we have eternal souls. Another problem is that most religionists seem to think we have no destiny past the grave except to join the type of God that they believe in. I believe that our destiny goes far beyond merely joining God; I believe we will continue "working" after we leave this life; and we will continue growing into something vastly different from what we are now.

So that we will be ready for the next level of existence, we must accumulate as much knowledge about creation, life, and other living beings as we can in one lifetime. We cannot reject any concept without thorough investigation. We must keep our minds open. A. Conan Doyle is reported to have said, "You have eyes and ears; use them." To that I add, "You have a mind; use it." We must be gullible enough to test new ideas, but skeptical enough to begin testing them.

This book attempts to bring scientists and religionists a little closer together. I hope it also brings the various religions together; although I realize that frequently is more difficult than convincing scientists to accept religious ideas.

Religions

There may be 10,000 more-or-less organized religions in the world, with maybe thousands of independent sects and cults. Most of them believe theirs is the only completely correct one and most of the other religions are teaching gibberish, trash, error, or evil. I believe most of these various beliefs contain *some* truth—and all of them contain either falsehood or only partial truth.

God wants to help humans find the right path to their destinies; and He continues to work with us. Unfortunately, almost all of us are overly concerned with our physical comfort and pleasure. These concerns are legitimate. God made us that way so we would take care of our bodies while our souls are within them; but most of us are more concerned about our present lives and less concerned about what God wants us to do. Even when we recognize that we are destined for a heavenly existence, many of us expect some kind of eternal life that is a carefree and pleasant existence vastly better but not too much different from what we would like to have on Earth. Some religions teach that our destiny is to join with God, without saying what that means or saying that we will lose our identity in that union.

My comparison of many religions made me believe the description of our destiny is much more complex than descriptions of either of those ultimate ends to our physical lives. Since I believe most religions contain some truth, I selected material that I think is at least partly compatible with teachings of many religions. Most of my comments are at least partly compatible with teachings of many religions, including Christianity.

The most important concept that humans should learn is a paraphrase of the first sentence in the second paragraph of the Declaration of Independence. "We hold these truths to be self-evident, that all men are created equal, that they are endowed by their Creator with certain inalienable rights, that among these are life, liberty, and the pursuit of their individual destinies." People cannot pursue their destinies if they are encumbered with fear, anger, hate, and pride. They must shake free of dominance by leaders and rulers that insist on blind obedience. The basic concept of Buddhism is start the quest for enlightenment by eliminating influence of previous teachings. Nothing that is written down is necessarily true. It is not necessary to believe something that is part of history and traditions. During the search for enlightenment it is not necessary to accept the lessons of teachers, because they might reflect personal opinions. All teachings must be tested. An idea or a teaching might seem correct, but still be wrong. Avoid fooling themselves because humans frequently are likely to accept ideas that are appealing to personal desires. Even so, people must have faith in themselves. In order to do that, it is necessary to recognize talents, skills, and interests.

THE AUTHOR'S VIEWPOINT

I am an engineer, not a theologian or politician; and I wrote this series of volumes more like a set of instructions for discovering our eternal destiny than as a theological or political dissertation. I tried to keep my writing in a straightforward manner with short words, sentences, and paragraphs, using a minimum amount of theological, technical, and foreign words.

I am a conservative/liberal in politics and religion, a centrist rather than a moderate. I believe in God. I believe the only begotten Son of God was incarnated in Jesus of Nazareth. I believe that the current state of human beings is the result of the process of evolution, which is the way that God created humans. While I believe many religions of the world contain great spiritual truths, I believe Christianity, even with its many divisions, is the only one that teaches it is possible for individuals to reach God independently of dogmatic religious leaders and allows individuals to grow spiritually at the rate that is consistent with their abilities. Since I believe in the teachings of Jesus Christ, I based most of my religious comments and references on that religion. Nevertheless, I recognize the great amount of contribution from the Jewish heritage of Christianity and from all the other religions of the world.

Not everybody that believes in a Creator believes that all humans have eternal souls. In other words, there is a God or an Intelligent Designer, but humans cannot exist in any form or manner outside of their bodies. Man is the product of evolution designed by the creator. This is the hypothesis of the book *Human Destiny*, by Pierre Lecomte du Nouy, an internationally recognized scientist. It was published in 1947. I read this book shortly after its publication and I have been slowly absorbing its message. Recently I read it again and I find that much of what I write here is similar to what is in that book, even though I began writing this book about a dozen years ago. I am thankful that I was introduced to the lessons of that book. However, I believe individual human souls can survive physical death and reside in a spiritual realm, something Dr. du Nouy evidently did not believe when he wrote his book.

Pierre Lecomte du Nouy demonstrated that the accident explanation for creation and the development of Mankind's present state could not be substantiated. He wrote that Mankind has a destiny to become supermen mentally and socially fit companions for the Creator; but individual humans that do not reach that level will simply perish. The importance of his book is in the brilliant way he brings evidence for creation of the material world and the intelligence behind physical evolution. This book may be what stimulated the concept of *Intelligent Design.*

Lee Strobel's book, *The Case for a Creator*, another book that helped firm up my beliefs, is somewhat like *Human Destiny*, except Strobel believes humans have souls. It presents evidence from the world of science to support the Christian's belief in God.

CREATION, EVOLUTION, REINCARNATION, AND OUR DESTINY

FORMS OF EXISTENCE

Your Existence

You can only have a destiny if you really do exist. Even if you exist, that does not prove you have an eternal destiny. However, recognizing that you exist is the starting point for discovering your destiny.

The fact of YOUR (MY) existence in some form is impossible to deny. If you did not exist, you could not be reading these words. However, philosophers and theologians have debated the form of existence during our entire historical period and probably for many centuries before that. Some philosophies claim the entire Universe is not real. Even if that is true, they have to admit that something like the Universe must exist (is real) in some manner, in some kind of place. The only thing any individual can know for certain is that he or she does exist—even though the form might not be completely or accurately known. It really does not matter to you whether you exist in a real world made of atoms and molecules or a place in your own imagination. What does matter is knowing what your place and purpose is, and if you really do have an eternal destiny.

When we try to determine our place and purpose in the world, we usually start with some assumptions, which may or may not be true. First, we really are independent individuals, each with our own consciousness. We have at least five physical senses: sight, sound, touch, smell, and taste. We have been told the Universe is composed of atoms, molecules, and other particles up to the size of stars that seem to be or act as if they were solid. Therefore, we assume the Universe is or acts as if all the visible objects are solid and do exist as they seem to exist. Then, we accept that there are billions of human beings on the Earth, all somewhat like one another. Last of all, we believe the Universe was created, either by accident or by some kind of creator. Based on these assumptions, we can start to analyze our place and our purpose for being where we are at this time and what we should be thinking and doing. The following list includes the starting points for the current study.

Believe you exist as an individual.

Believe that your physical senses are all you need to understand the world in which you live.

Believe the Universe, including the Earth, is what it seems to your physical senses.

Accept that many forms of knowledge are available to you from many sources.

Understand and accept the amount of knowledge you are capable of absorbing. Try to expand your ability to understand greater amounts of knowledge. Resist the urges of demagogues (usually some kind of religious or political leader) to generate hate, prejudice, and bias—all factors that limit your ability to acquire and utilize true knowledge.

Following the example of Buddhists, eliminate the influence of previous teachings. Nothing that is written down or told to you is necessarily true. It is not necessary to believe something that is part of history and traditions. It is not necessary to accept the lessons of teachers, the haranguing of politicians, or the ranting of demagogues because they probably reflect personal opinions. The knowledge in these sources might be true; but you should test the truth and significance of all teachings. An idea or a teaching might seem correct, but still be wrong, or at least, partly wrong. Avoid fooling yourself; because humans frequently accept ideas that are appealing to personal desires. Even so, you must have faith in your self. In order to do that, it is necessary to recognize your talents, skills, and interests.

Realize that you may see things differently than I or others do. I believe in a Supreme Being that created the Universe and all life forms. You might not agree with me or anything else I say in this book or in any other thing that I have written. However, do not discard my comments without completing this book and comparing my thoughts with what you have learned previously and what you believe about those things. You might still not agree with me, but I hope you will be closer to God or at least to self-understanding because of renewed study of your self and whatever document(s) in which you have faith. Pray or meditate before making important decisions, which includes decisions about how to act, what faith you accept, and what rituals you think are important.

The next thing we must decide is whether we must follow some celestial plan for our lives or if we can make completely independent decisions about our lives and our modes of living. Either we are predestined to perform certain actions and reach a predetermined end, or we have freedom to control our lives and our futures.

Predestination

If your life is **Predestined** (predetermined), there has to be a God or some other kind of creator that established your destiny. You are unable to alter your life or your circumstances from what has been determined for you. Your thoughts, feelings, concerns, actions, and everything that you think are the results of your own mind are actually the results of a plan previously decided for you. Even if you think you are rebelling against your fate, your rebellion is part of your destiny. You cannot avoid whatever happens to you. There is some second in the future when you will die physically, and what happens after that cannot be changed.

Whether you are good or bad is part of the plan; and you are not responsible for your behavior or character. If you could deviate from this pattern, the entire future would be altered. Even the slightest alteration in your life would affect everything in the world, if not immediately, eventually. If, for example, you were predestined to meet a future spouse at some certain place at a certain time and if you *could* fail and did fail to go to that place, you would not meet the person intended as your spouse. In addition to being at the proper place, you must also fall in love and decide to marry the person. The other person must respond to you in the proper way and agree to marry you, and so on and so on. Any alteration in your predestined affairs would change your future life and the lives and destinies of the intended spouse and future offspring possibly resulting from that union. Not only would your actions affect your intended spouse and offspring, but everyone related to both of you in any way would also have the plan for their lives altered. Eventually this would alter the entire celestial scheme of things for everyone.

Predestination seems incompatible with everything taught about good and evil by traditional religions, even those that teach belief in predestination. You should be rewarded for being and doing whatever you were supposed to be or do, whatever that might be. If you are predestined to live as you do, you should not be blamed, nor should you be punished for doing what you are supposed to do, regardless of how evil it might be. You will die when your "number is up," no matter what risks you take. Therefore, you can do whatever you think you are "supposed" to do. This concept eliminates all challenges to an individual; it denies the possibility of growth based on personal efforts; and it describes a meaningless existence—from the individual's point of view. You are merely an actor on a grand celestial stage blindly following the script and the guidance of a celestial stage director.

This concept is similar to the fundamental belief of Gnosticism. There are many varieties of beliefs included in the term Gnosticism; but they all share the belief that some people are destined for salvation and some are destined for destruction. Only certain people are capable of knowing and understanding the secrets of God and being saved. Others who cannot acquire that knowledge will not be saved. Salvation comes through knowledge, not works or faith.

Complete Free Will

You might have **Free Will** and be completely free without a predetermined destiny. You are in complete control of your life; and you have to devise your own purpose for living; it is entirely as you want it. God, if you choose to believe there is such a being, is not concerned about you or any other human, except perhaps as an interested observer. You do not have a predetermined purpose in life. You are completely free, with all your actions strictly the result of your own thoughts and desires. You are free to accept or strive to change what you feel are restrictions on your life. You might wish you were of the opposite sex or a member of a different racial or ethnic group, or wish you were born richer or better looking or in another country. In modern times, at least in the USA, many of these conditions can be overcome with patience, skill, effort, money, and/or some other kind of sacrifice. In this case, there will be some natural restrictions on your complete freedom. Obviously, you cannot defy natural laws such as the pull of gravity, the flow of time, or results of carelessness, nor should you violate civil or criminal laws.

As in lives of most humans, the world will probably be unaware of your existence and what you do or think will not affect the future of more than a few other people. Your goals may or may not be what others think you should have. That is irrelevant. You are the *"master of your fate, the captain of your soul."* (From *Invictus*, by William Ernest Henley.) This complete freedom can be either exhilarating, or frustrating. You cannot blame God or other humans for your failings or your troubles. Even when others accidentally, incidentally, or deliberately cause you harm or interfere with your plans, the way you respond is up to you. Many things that happen to you are outside of your control and some of them might harm you. However, sometimes, these misfortunes can be adapted to your plans; or you might be able to modify your plans, and eventually you might receive some kind of benefit from the misfortune. This does not necessarily mean there is no Creator or Supreme Being. The Creator might not be interested in you; or He might have some reason for allowing you to live in any way you want to.

Atheists and Anti-theists believe in this view of personal existence. Agnostics and some "modern" religious leaders also believe in this concept.

Free Will with a Purpose

There is a third possibility, which is a combination of the two preceding possible scenarios: you could have **Free Will with a Purpose.** You still have a predetermined purpose and destiny; but the way you reach that destiny is up to you. Even though you have a pre-determined destiny, you can deviate from the plan. There may be celestial forces from time-to-time that direct or encourage you into a predetermined pattern. In other words, you might be partly free and partly predestined. Even though you have a pre-determined destiny, you can deviate from the plan if you want to. The way you act and think and any consequences of your decisions are up to you. Since you have free will, you can choose whether to believe you are merely an accident of nature without an eternal destiny, or you were created for some purpose and you have some kind of destiny.

There may be many ways to get to your "destination" with free will. Here again, if you deviate from the predestined pattern, others will be affected. Every time that you act in opposition to the celestial plan, you could be upsetting the plan or the destiny for yourself and at least a few other people. We assume that the people that are deflected from their original destiny by your actions or failures to act will have another plan prepared for them. God is a real person to you and sincerely interested in the choices you make. This God will try to help you reach your destiny; but you always have the freedom to accept this direction or refuse it.

It's Not Your Fault

People anxious to justify their own behavior or to sell some "miracle" product use the phrase, "It's not your fault" that you are the way you are. Whether the problem is being overweight, drinking, smoking, homosexuality, abusing children, any kind of criminal activity, or dozens of other problems that interfere with spiritual development, it's not your fault—it's caused by your genes or your environment. If you can't read, write, or do arithmetic, it's not your fault—the government did not provide enough money for your education. If you follow oppressive or dishonorable business practices, you do it because customers demand lower prices. Unsympathetic spouses force you into adultery. You are a criminal because you are too poor to take care of yourself or other obligations. You are in poverty because the government did not provide enough money. You say, "God made me this way! I can't help it," or "The devil made me do it."

You were born in the body and conditions that you are supposed to be in. You have to live with what you are and have. You are supposed to do the best you can with that body and under those conditions. It is up to you to overcome weaknesses and problems in those conditions. You will be judged by how well you did with what you were given. Your spiritual advancement depends on how much you advance during life on Earth. Jesus expressed this attitude toward personal achievement quite clearly in the following passage from Luke.

> **Luke 12:47** And that servant, which knew his lord's will, and prepared not himself, neither did according to his will, shall be beaten with many stripes. But he that knew not, and did commit things worthy of stripes, shall be beaten with few stripes. For unto whomsoever much is given, of him shall be much required: and to whom men have committed much, of him they will ask the more.

God gave us the Commandments to tell us what we should do. Since God gave them to us, it must be possible to follow them—and not to ignore them or violate them. These rules were not merely to bring glory to God. Following these rules is intended to help us overcome our weaknesses and grow spiritually—bringing us closer to God in the process. Our entire lives should be a continuing process of learning and improving our selves. Every religion encourages individuals to improve levels of knowledge and to learn how to love their neighbors. The difference between religions is in the kind of knowledge that is encouraged and in the definition of the word "neighbor."

Solomon, described as the wisest man that ever lived, discussed the value of the right kind of knowledge.

> **Proverbs 1:1-7** The proverbs of Solomon the son of David, king of Israel; To know wisdom and instruction; to perceive the words of understanding; To receive the instruction of wisdom, justice, and judgment, and equity; To give subtilty to the simple, to the young man knowledge and discretion. A wise man will hear, and will increase learning; and a man of understanding shall attain unto wise counsels:
>
> To understand a proverb, and the interpretation; the words of the wise, and their dark sayings.
>
> The fear of the LORD is the beginning of knowledge: but fools despise wisdom and instruction.

I believe humans have free will with a purpose. The purpose is to do the best we can under the circumstances of our lives. The Commandments and/or other rules within every religion are meaningful only if we have complete freedom or free will. There would be no reason to give rules to people that could not deviate from the rules.

Therefore, your errors and misdeeds *are* your fault, at least most of them, are your fault. Regardless of who is to blame, there will be consequences for failure to grow spiritually. We must do the best with what we have.

The Real World

Consider the proposition that you are the only being in existence. This is worth exploring, even though it violates our initial assumptions. You have to start somewhere and the only thing you know for sure is that you exist. If you are the only being that exists, then you are the creator of all that is perceived by your senses. Even the senses that you use to perceive with are your creation. If you are the only being in existence, you created the world. You cannot have free will because you made the world the way it is and everything that happens to you is the product of your plan. You are predestined, you made it that way; and whatever you do or think cannot have any effect on other people, because there are no other people. You are riding an amusement park roller coaster that is on a fixed track and will go where you planned it to go. Whether or not your actions affect your future is another problem. Of course, just because you are alone in what you call this Universe, does not mean you are the only being that exists anywhere in any world. You might be a member of some group of beings just like you in another realm, or in what some have called a "parallel Universe."

Why would you choose to invent such a condition? Perhaps you chose to amuse yourself, perhaps to develop your abilities, perhaps because you were lonely, or perhaps because you became insane. Then, everything that happens to you is caused by your own mind, whatever that mind may be. You suffer losses, become ill, or are in pain to prove that you are "real." You fall in love to prove that you are not alone (paraphrasing C. S. Lewis).

Even though you are the only being in this world, you might have been placed here by some outside force. If some other force has allowed, placed, condemned, exiled, or otherwise caused you to be in this condition, why or how did this occur? We can come up with some answers to "why," but the answer to "how" is completely beyond our scope. The "why" might be the same as why you would choose to create this world by yourself with the addition of you are being rewarded or punished for something you did in this other world. Maybe where you are now is what the Roman Catholics call purgatory. Maybe this is Hell. I rather doubt that this is Heaven, although it might be.

If YOU can exist, it is reasonable to expect others to exist in the world of your physical senses, in your dream world, or in the "place" where you really belong. If these other beings coexist in the same dream world with you, they are just as "real" as you are. Then, these others share in the creation of this dream. If they exist anywhere, they are either predestined or they are free and they have the same freedoms and/or restrictions that you have. (I am not going any further in this direction. It may be interesting, but this does not advance what I intend to discuss.)

Even more perplexing is the definition of "human nature." Is human nature unique to humans or is it similar (at least in some ways) to the nature of other living creatures, including vegetation and microscopic material sharing our world? Is human nature part of our evolutionary heritage, or is it continually being recreated as cultures develop and change? Perhaps human nature is somewhat like the nature of the Creator—if there is one. These questions are part of the reason we do not understand what is real and what is not. The problem with knowing about the "real" existence of the world is the same as knowing about anything. Before going much further, the functions and capabilities of human beings will be discussed.

PERSONALITY

Definition of Personality

Regardless of whether you are free or predestined, you exist as a personality. The word "personality" has two meanings which are closely related to one another but have different connotations.

How you appear: This is the way others perceive you. You may act aggressively, shyly, friendly, humbly, religiously, angrily, meanly, etc. You may appear to have a dynamic, charming, charismatic, pleasing, nasty, etc. personality. It is based on your attitude toward others, and your opinion of yourself. Your actions are usually determined in some degree by the people you are with. Most people act differently with different people: family members, male or female acquaintances, romantic friends, younger or older people, authorities, and so on; therefore, different people might think a person has a different personality than others think he or she has.

Who you really are: This is self-identity. It is the ego. This is the permanent part of an individual; but it changes during life on Earth, depending on environment, personal relationships, other personal experiences, and the reactions to those conditions. However, few people ever change their personalities very much.

Psychologists do not agree on the factors that cause or contribute to the origin of individual personalities. Environment versus heredity is a continuing argument, one feature gaining popularity, then the other, depending on what form of political correctness is in vogue at any given time.

Many factors might affect the creation and development of a personality. Most of these are common to everyone and probably most have contributed to a person's present condition. They are listed below in pretty much the chronological order they affect people, with each condition and experience influencing the degree of influence of subsequent events. Some of the factors are based on ancestry and circumstances of infancy and are almost entirely unavoidable. Others are repetitive and periodically continue to influence the further development of what you are. Most of the factors can be modified at any time with sufficient desire and encouragement and assistance from others.

Reincarnation: if there is such a thing.

Genetic memory: if there is such a thing.

Heredity: the physical characteristics, including racial features that have been transmitted to you through the genes that were acquired during fertilization.

Pre-natal conditions and experiences: the mother's health, habits, diet, injuries, allergies, etc.

Your body: sex, physique, genetic handicaps, if any, right or left handed or ambidextrous, and other unique conditions that influence strength, mobility, and all inter-personal activities.

Early training: family or social traditions, legends, prejudices of parents and relatives, potty training, walking, and other pre-school actions.

Physical problems: Childhood diseases and injuries.

Status: social and economic.

Treatment and discipline: from parents, other adults, siblings, baby-sitters, teachers, etc. during childhood.

Parents: whether both parents are present and active in the life of children, and other relationships with parent(s).

Siblings: relationships or lack of with siblings, and order of birth—first, middle, or last-born.

Other relatives: relationships or lack of with other relatives.

Other people: relationships or lack of with other children and adults, especially teachers and fellow students.

Religious training and experiences, if any.

Authorities: relationships or lack of with teachers, police, and other authorities.

Formal education: type, amount, and method of teaching.

Other childhood experiences.

Personal spiritual experiences, if any.

Employment: history or lack of, and what kind of work.

Matrimony: being single, married, separated by divorce or death, and at what age.

Children: how many and when they were born, if any.

Responsibilities: or lack of.

Traumatic experiences: illnesses, accidents that result in long-term or permanent disability, etc.
Chronic ailments: allergies, and other physical conditions.
Natural changes: maturing and getting old.

Four kinds of personalities

There are four distinct types of personalities. Many people are combinations of two or more of the four kinds, sometimes with none of them dominant. The degree of intensity or impact varies with each individual and with each change of circumstances and age of the individual.

Intellectual: craves freedom, knowledge, and wisdom.
Mystical: loves spiritual and/or religious studies and meditations.
Passionate: demands rules, power, and fame, needs emotional stimulations.
Sensual: seeks pleasure, enjoys ritual, wealth, and luxury.

If you are predestined, your personality is the way it is "supposed" to be. If you are completely free or have free will, you are responsible for the way you respond to the factors of development and eventually what your personality actually is at any time. Whatever the source of it, the most obvious part of your personality is your body. The physical body and the way you behave is an extension or expression of your mind. (Books are available that describe "body language" referring to the way a person's body reflects his or her thoughts.) Certainly, your gender, size, appearance, health or lack of health, and chronological age have a major impact on how you appear to others; but your mind still has control over many aspects of your appearance, health, and effective age. Your attitude, aptitude, and discipline have major impacts on the condition of your mind.

Attitude

Attitude is defined as the mental reaction toward facts, events, or people surrounding or affecting an individual. Attitudes strongly affect the way people look at authorities in scholastic, employment, social, political, or religious environments and the way those people see a person. Attitudes range from positive to negative. People should try to understand what their attitudes are; and harmful ones should be modified to help a person be more productive in personal relationships and be more productive in professional and spiritual activities.

The following list includes attitudes that reflect an individual's personality and could affect further development of an individual's personality. While some of them are not compatible with other attitudes, a person might have more than one of some listed attitudes. For example, a person might be radical and negative, or liberal, negative, and pessimistic.

Skeptical: believes truth is relative and uncertain. Doubts all new and unfamiliar ideas and reports of events not personally acquainted with. Distrusts other people.

Radical: tries to make extreme changes in existing views, social conditions, and government or other institutions of his or her society. Hate and anger are easily aroused. Participates in demonstrations and violent actions. The degree and amount of changes demanded become less and less as one gets older.

Liberal: freely accepts and learns new and unusual ideas. Willing to give money, time, and efforts freely to a cause felt to be worthy. Prefers change to stability. Reluctant to accept traditional beliefs, rejects authority, and resents having to conform to traditional customs and beliefs. Searches for injustices: real, imaginary, or exaggerated.

Positive: believes that trials and disasters can be overcome. Believes many problems that cannot be solved can be adapted to fit personal needs or desires. Tends to be active and impetuous. The expression, "If you are given a lemon, make lemonade" reflects positive thinking.

Optimistic: believes that things will usually turn out right, regardless of what is happening at the moment. Sometimes this attitude is a result of laziness.

Open-minded: receptive to arguments and new ideas. Not committed to any particular philosophy or cause until they have been tested and accepted. This is not necessarily a harmful attitude, but an open-mind might merely be an empty mind.

Conservative: reluctant to learn new and unusual ideas. Willing to give money, time, and efforts freely to a cause felt to be worthy. Accepts, or at least tolerates, traditional beliefs and authority. Prefers stability to frequent changes. Emotions not easily aroused. Not inclined to take action in support of accepted ideas; moved only by serious threats to status quo.

Realist: believes that some things can be changed, but not everything, and that some things will turn out right and some will turn out to be harmful. Believes that a person should do what is reasonable and appropriate for existing circumstances.

Pessimistic: believes that things will usually turn out to be harmful. Sometimes this attitude is a result of laziness and is an excuse for not doing anything to solve problems. Extreme pessimistic attitudes merge with negative attitudes.

Negative: believes that nothing ever turns out right. If it appears that things seem to be going well, eventually something bad will happen to cause difficulty or pain. Tends to be passive and lets other people do what is necessary. In severe cases, negativity can result in physical of emotional disturbances.

Cynical: fault-finding and critical. Believes humans are motivated only by self-interest: greed and selfishness. The cynic meets his or her own definition, being concerned only with his or her own beliefs.

Discipline

Discipline is defined as training that corrects, develops, or perfects mental abilities and thinking. This includes the ability and desire to learn and develop control over ones desires, attitudes, and needs. The degree of discipline acceptable to a person is demonstrated by willingness to follow rules established by society and acceptable authorities. The following list includes the most common responses individuals make toward discipline. Here again, it is possible for a person to have more than one response to discipline. For example, a person might be antagonistic and gullible at the same time. This happens in religious, social, and political areas if one authority is hated, and some other authority expresses the same hatred. They would work against the hated one because they blindly accept the teaching of the other.

Antagonistic: actively expresses opposition to authority. Always angry. Selfish. Conceited. Inclined to take violent or destructive action. Works vigorously to gather others with antagonism toward the same causes or authority.

Rebellious: resents existing rules of society. Accepts but does not seek membership in social organizations, but joins with other rebels in specific causes.

Independent: refuses to accept ideas and rules that do not seem right. Tends to resist joining organizations, especially secret societies and fraternities and sororities.

Sympathetic: recognizes interdependence of individuals. Does not argue with authorities or create discord. Likely to join groups or organizations that provide a sense of security.

Submissive: accepts almost every demand of society and recognized authorities. Joins groups and organizations because of social pressure.

Gullible: easily cheated or convinced. Follows anyone that speaks authoritatively, regardless of merit or need.

The most valuable combination of attitudes and discipline traits is being: *realistic, open-minded, and independent.* These positions provide the greatest opportunity for development and improvement of a personality.

THE BRAIN, MIND, AND SOUL

THE BRAIN

The brain is purely physical. Animals, including insects and most other living beings have brains of varying sizes and complexities. It is located in the skull; but the nerves attached to it are in almost all parts of the body. Eleven parts of the brain have names to help physicians and psychiatrists understand human beings; but those names are not important to this discussion. However, the brain does have a limiting impact on the functions of the mind, which is important. The brain grows in size and complexity as the body grows and matures, providing more capability for the mind to grow. The brain grows whether the mind does or not. It also can degenerate because of diseases, drugs, or accidents. Degeneration of the brain limits the effectiveness of the mind; because the brain links an individual with the rest of the world.

Basic instincts such as fear of falling, fear of loud noises, hunger for food, water, and air, sex impulses in humans and animals, how birds know how to build nests, or dogs know which animals they should pursue and are capable of defeating seem to be built into the physical brain; because these functions do not have to be learned. Instincts seem to be associated with independent movements of the body; although the conscious mind of humans can resist most of them. The conscience might also be part of the brain, and be part of basic instincts.

The Physical Senses

The brain controls the physical body through the nervous system. The sensations that bring us into communication with the outside world (outside of US as individuals) are interpreted by the brain, or mind to give us an understanding of what is occurring around us. Our senses are all really one sense: the sense of touch, or the response of nerves to stimulation. The eye is a camera that concentrates light onto a group of nerves that respond to the touch of light vibrations. The ear contains an external drum that responds to the touch of vibrations of certain frequencies, concentrating the responses into the inner ear where specialized nerves of various lengths tuned to individual frequencies react, sending signals to the brain. The nose is a concentrator of scents and odors that pass through the nostrils and touch nerves sensitive to certain gaseous or vaporous stimulants. Various nerves in the mouth and especially the tongue respond to the various touches of saltiness, sweetness, bitterness and sourness. All of the skin, especially on the hands and feet, responds directly to contact with various stimuli. Inside the body are other nerves that respond to bodily functions and malfunctions. (Sensitivity to all outside stimuli can be improved with need, training, and practice.) The brain interprets these nerve contacts as sights, sounds, odors, tastes, and as various kinds of textures, temperatures, and other sources of external stimuli depending on which area of the brain is involved.

Our senses respond to changes in stimuli; and they are limited in their responses by the frequency of the stimuli. If there were disturbances other than what we detect with the five senses, at frequencies lower or higher than the various nerves are capable of responding to, we would never know what they are through the five senses; unless the frequencies outside of our range affect something that translates them to frequencies in our detectable range. For example, X-rays can affect photographic film, allowing us to see their effect. Magnification of objects or areas of space can overcome the limitations of the eyes to detect signals too small or too close together for the limited number of nerves in the retina.

In addition to the five, seemingly separate senses, are other responses to more esoteric things. When we say we are feeling good or feeling bad, we are using some unnamed sense of understanding of our physical and mental condition. Feeling pain is more complicated; because pains resulting from infections and other disturbances within the body seem to be transferred to the outside. For example, being dehydrated can cause pain in the forepart of the head; simply drinking two glasses of water usually can cure headaches resulting from dehydration. When people have paranormal experiences they are probably using some other sense—popularly called the "sixth" sense. However, there might be more than six senses.

The brain is not the mind. The brain responds to physical senses and under some circumstances causes the body to react with some kind of movement. The mind usually can override the body's tendency to respond to physical stimuli. For example, when the brain recognizes the body itches somewhere, the brain's first response is to scratch where it itches; but the mind can restrict the physical attempt to relieve the problem.

The mind's relation to the brain is like the programs that make a computer function, such as Windows and its subsidiary programs like word processing features. Nevertheless, the brain has a limiting effect on the mind. The mind develops and uses the brain for storage of information and learned behavior. Diseases, chemical abuse such as with alcohol and other drugs, and just getting older reduce the ability of the brain and the body to react properly to the mind. The effect of simply getting older past middle age in absence of disease causing problems with memories may be because the memory "bank" becomes overloaded and retrieval of specific information is slower because there are so many bits of information to be sorted through.

Being conscious does not necessarily mean being aware of what is going on around a person. Sleepwalking, automatic writing, being hypnotized, and a few other activities that seem to be occurring while not completely aware of what people are doing are examples of being conscious but not aware. Once, while I was driving after a long day of driving, I suddenly realized I was about a mile further than I was the last time that I was aware of being on that road. I decided I needed to stop and get some coffee. A similar thing happened once while I was in the Navy on guard duty during the small hours of the night: I was walking my rounds and suddenly I was about 100 yards further than the last time I remembered. During these times, I must have been driving and walking in my sleep.

Automatic writing is a good example. Spiritualists believe a deceased person working through the writer controls automatic writing. However, whether this is the case or not, the body performs a task without awareness of the writer. Somehow, the person is able to surrender control of the body to the subconscious mind, which seems to be acting independently of the writer.

On the other hand, some people who are unconscious or in comas, frequently are aware of what is happening around them. A friend of mine who was in the last stages of dementia seemingly was unconscious; but he began to weep when I read some of his favorite Bible verses to him, which indicated he was aware of what I was saying; and he reacted in this manner. Betty, my wife, told me that her late husband was in a coma and seemed to be completely "out of it." His brain was in an advanced stage of physical degeneration. However, a nurse told Betty that while she was visiting him in the hospital, his blood pressure increased, indicating he was aware of her presence.

Later in this book, I discuss the study made by Doctor Raymond A. Moody. Jr. about people, who seemingly were clinically dead and yet when resuscitated, relayed stories about experiences that indicated their minds or some part of their minds was aware of the experience, and they remembered it. Many people reported similar experiences, indicating that the experiences were not unique to a single individual. These reports seem to be reporting a spiritual or psychic condition.

These various experiences demonstrate an ability of the brain and mind to operate independently of one another. Therefore, I will continue my discussion assuming that the two are separate parts of an individual personality.

THE MIND

Your mind determines your personality. The mind is the source of all your thoughts and emotional reactions to outside stimuli; and it controls the way you think, behave, and, pretty much the way you appear to other people. Your thoughts are apparent in your face and through your body language. Thoughts can create and aggravate many physical ailments, directly and indirectly by choice of lifestyle: foods eaten, choice of associations with other people, and reactions to temptations. How you react to others and things that happen to you determines what you do with your life and what happens to you in your present life; and it is the most important factor in determining and reaching your eternal destiny.

The Parts of the Mind

The mind has four functions that act as if they originated in separate parts of the mind: *the conscious, subconscious, unconscious and hyperconscious*. Most experts in the fields of study about this subject do not accept existence of these four as separate activities using separate parts of the mind. However, the mind acts as if they do exist. Therefore, I use these terms.

The Conscious Mind controls deliberate movements of the body, words, and other voluntary actions of a human being's body. The conscious mind receives, and reacts to the five senses and other sensations originating in various parts of the body and the total mind, causing voluntary actions to take place or restrict them from acting under some circumstances.

The Subconscious Mind stores, processes, and retrieves memories of conscious thoughts, and events observed or detected by the conscious mind. It acts as an intermediary between the conscious mind and the hyperconscious mind and searches for long-lost memories and contributes to creative processes.

The Unconscious Mind is hidden, but still quite active. It is formed during the gestation period; but it changes during life according to training and other experiences. It manages the automatic activities of the body such as controlling the heart rate, keeping balance while standing and performing other routine movements, controlling the flow of hormones, and managing the respiratory system. This area controls acquired automatic functions such as walking, talking, using a keyboard, hitting a golf ball, driving a car, and similar abilities that change or improve with learning, repetition, and discipline. For example, children taught to use a musical instrument before the age of twelve have unique growth in parts of the brain. This particular form of growth probably is where these learned skills are stored and is part of the unconscious mind.

This part of the mind is probably the source of psychosomatic problems, habits, allergies, addictions, and our likes and dislikes. This mind can be fooled by tricks such as placebos and hypnotism. It has a very powerful influence on the conscious mind and can be helpful or destructive. The unconscious mind is something like the basic programs built into a computer. Once a program is installed in a computer, it takes special actions to uninstall it; and sometimes it takes an expert to modify or eliminate it. The admonition to think positively is aimed at influencing actions of the unconscious mind that interfere with pleasant living. However, thinking positively is not enough—it is vital that this mind should not have any *negative* thoughts. Jesus mentioned this conflict in one of his sermons to his disciples. He was talking about prayer; but that is the same as any other kind of thinking.

> **Mark 11:**23 For verily I say unto you, That whosoever shall say unto this mountain, Be thou removed, and be thou cast into the sea; and **shall not doubt in his heart**, but shall believe that those things which he saith shall come to pass; he shall have whatsoever he saith.

Activities that are normally controlled by the unconscious mind are influenced by the conscious mind when it is active. An example is the act of breathing. One way to tell if a person is sleeping is to watch the breathing. If it is regular, the person is probably asleep, because the unconscious mind is in complete control. During waking hours, the conscious mind causes breathing to fluctuate for various reasons, for examples: sighing in relief, to clear ones nostrils, or hyperventilating in panic. Hyperventilating is a peculiar condition—by flooding the brain with an excess of oxygen, a person will faint, allowing the more reasonable unconscious mind to take over the act of regular breathing. A similar situation exists when performing an acquired habitual activity such as golf, tennis, or bowling. If the conscious mind is trying to perform the appropriate movements and reactions, the results are usually disastrous. For example, it is not unusual for a player that is close to a record to fail in the last play or throw because the conscious mind is trying too hard to control movements that should be performed unconsciously. People with serious physical injuries sometimes are kept in an unconscious state to prevent the conscious mind from causing movements or thoughts that could prevent or slow healing of the body.

While asleep, the conscious mind relaxes, relinquishing control of the body and subconscious mind. The unconscious mind will still receive information through the physical senses and cause the conscious mind to regain control if the stimulus is strong enough. If the amount of light suddenly increases or decreases, the body will usually awaken. Sudden loud noises will usually awaken a person. The unconscious mind's primary function is to protect the body, including the brain.

The conscience acts through (or from) this part of a person's mind. It brings information stored within it to the conscious mind. In this function, it acts like the automatic spelling and grammar checking in the word processing part of a computer. When a person does or thinks about something that can be harmful, the conscience automatically advises the conscious mind of the probable effects. Either it can allow the senses and sensations to dominate actions or it can follow directions of the conscience and override demands of the senses.

The Hyperconscious Mind is the most important part of the mind: it is also called "The Soul." It will be discussed in the following section titled *The Soul.*

The Memory

The memory is a product of all parts of the mind and seems to have five stages of recording information.

A very short-term memory of less than a minute, probably no longer than a few seconds. Incidents and thoughts recorded here exist only in the conscious mind. If something happens that causes a person to lose consciousness before information within the short-term memory is transferred to the slightly longer-term memory, the material is lost. This is similar to what I am typing right now. If the computer is turned off before it is saved, it will be lost.

In addition, if some action or thought is not considered important by the mind, it is lost shortly afterwards. When we try to remember if we have performed a routine task, we cannot remember—at least not for sure. Probably all of us have experienced the confusion of forgetting if we have done some routine task, such as locking a door, turning off the gas, or whether we took our medicine on schedule. Another example is when a person is counting something in sequence, such as taking a pulse, the numbers are recalled as they fit in the sequence; but sometimes, after lower numbers are used, it is difficult to be sure previous numbers were actually considered.

A slightly longer-term memory contains items acquired and "saved" during the previous few minutes if the memories are reinforced somehow, or related to some familiar object. Otherwise, they might be lost. A common example is trying to remember the name of someone we met during routine experiences, such as at a big party or during ordinary business or social meetings. However, if that person or the name can be associated with something that has been recorded earlier, it can usually be recalled, but not always immediately. Frequently, when trying to remember someone's name, we get the first letter right, such as thinking Snyder, when it should be Simpson. This is probably is because we associated the new name with the other name and they are "filed" in alphabetical sequences.

A middle-term memory contains items transferred from the short-term and the slightly longer-term memory after being reinforced in some manner, such as combining the new item with others previously recorded. Items in this memory are retained until a person has enough dreams to process them.

A long-term memory contains memories transferred from the middle memory where it was retained for no more than a few days. During sleep, dreaming seems to act like compressing files (or defragmenting) in a computer, making the actions of the day more orderly and less disturbing.

While dreaming, experiences and information acquired during the daytime and occasionally within a few days earlier are compared with previously stored data within the brain. If the new information seems important for some reason and does not create unpleasant memories, it is stored in this long-term memory area, and is readily available when it is required. If new information is associated with some experience or person that we consider unpleasant forbidden, or undesirable, we discard the memory. For example, a man I knew who had an excellent memory seemed unable to remember his secretary's name, which was also his wife's name. I don't know if he disliked his wife or if the name stirred some other feeling he had for his wife. If unpleasant experiences have been severe or repeated frequently, such as wartime battles or other frightening situations, or if we have done something we should not have done, the experiences will be recalled during sleep many times; because they never have been properly resolved and eliminated.

Dreams also help us process frustrations, conflicts, and problems acquired during the day. Similar situations or experiences from the past are compared with the new experiences; if they were resolved somehow, we are assured that the new ones also can be resolved. If the previous ones were not completely resolved, our dream might be recalled as a nightmare when we awaken. Nevertheless, even nightmares might help us understand why the previous situations were not resolved and present us with possible solutions. A frequent dream I have is of being in school and trying to get to a new class on time or else I cannot find the correct room. I think that dream occurs whenever I am confronted with a new situation of any kind. It reminds me that I survived all my classes in school and I can survive this new situation.

A dead-storage memory contains the material discarded during dreaming or not used during the previous month or so and will be retained indefinitely deep in the mind. This fifth stage is like computer information that is sent to the recycle bin for compressing. This information is probably stored in the hyperconscious mind.

Within the subconscious mind is a coordinating or integrating function that searches for memories in the dead storage memory in order to bring them to the conscious mind. Searching for a specific memory can take a long time. For example, trying to remember the name of someone not seen in a long time while talking to them may be unsuccessful; but sometime later, maybe days later, the name will suddenly appear in the conscious mind.

Dreams

Dreaming is a special function of the mind. Scientists have been researching the phenomenon of sleep, which included study of the process of dreaming. It seems to be more important to dream than to get a good night's sleep. I read about a study a long time ago that demonstrated this. In the study, volunteers were not allowed to sleep for days until they could not stay awake any longer. It was expected that they would go into deep sleeping modes; but they did not. They spent a long time in the less deep sleep where dreams take place.

While dreaming, information can be transferred from the hyperconscious mind to the subconscious, where it can reach the conscious mind when waking up. These thoughts may be the source of sudden understanding of or the solution to a problem. They may also be the source of information about spiritual subjects. However, for a dream to be available for use by the conscious mind it should be recalled immediately on waking or it could quickly be lost. If the dream seems significant, it is best to write it down before getting out of bed or doing any other task.

The noisy wheel bearing: Many people have reported discovering solutions to problems that had worried them during waking hours. I had one that is typical of some reports. We had an automobile that started making a horrible noise. I did not know much about autos then, so I could not find the cause. A new baby in the house demanded all of our spare money and I did not want to spend a lot of money on the car. So, I just worried about the noise. One night in a dream, I was on a road waiting for a train to pass. As it approached me, I heard an awful noise, like the noise from my automobile. I mentioned it to a man standing near me, wondering what could be causing the racket. He said, "It's that front left wheel. It must have a bad bearing." When I woke in the morning, I told my wife about the dream, hoping one of us would remember it. As soon as I had an opportunity to look at the front left wheel bearing of my car, I checked the truth of dream. Sure enough, that was the problem, which I was able to remedy by myself at little cost. In cases like this, it seems that the subconscious can understand or know about things not consciously known.

The Noisy Airplane: Another dream I had was a different kind—the sudden recall of an experience in the past that seemed to be happening in the present. I was deep in a nap on a couch with smooth fabric covering, when the sound of an airplane close above my home partly awakened me. In my half-awake condition, I was taken back to a time when I was about nine years old, many years prior. I was back at that time and I was lying on the couch in the front room of our home when the sound of an airplane close above my home wakened me. I could feel the mohair upholstery on which I was lying. I could see the room with a black pot-bellied stove near the wall opposite from me. I saw the curtains on the window. The dream was very real. It was more like a waking experience. I wondered where the rest of my family was. Whatever it was, the dream was very real.

Dreams like this can be caused by noises or other stimuli that tend to awaken sleepers. The mind attempts to keep the body sleeping and pretends that the stimuli actually are parts of a dream.

The comforting presence: Another dream that seemed to be real was of the sudden presence of my wife who had died several months prior to this dream. I dreamed that she appeared in the doorway to my bedroom. I was startled and told her I thought she was dead. She replied that she was, but had come back to see how I was getting along. I said I was okay—I could cook well enough and felt fine. She asked me if I was doing the laundry. After I stated that I was, she said, "Well then, there's one more thing." She got into the bed beside me and we hugged. I dreamed that I woke up and she was gone, leaving me with only my pillow. In my dream, I wept profusely, saturating the pillow. Then I woke for real. There were only a few tears in my eyes and I was alone. I felt a little better after that. This dream was probably a wish fulfillment. Of course, it might have been a problem-solving dream, telling me I was going to be okay. The dream, at least partially, solved the problem of my grief about losing her. Perhaps my wife's spirit actually was there in bed with me, trying to console me. Maybe she was worried about me and returned to comfort herself?

Woman in prison: Another dream that seemed real was about a close friend who had been in a coma for several months. During that time, she had several setbacks and recovery was hopeless. I went to visit her and prayed for her recovery. Two times when I did this, she turned her head very slightly away from me. I realized that she wanted to be freed from life. I assumed that was because she must have known she would never regain normal existence. The last time I visited her, as I prayed, I asked the Lord to release her from her suffering, without asking for healing. That time her head slightly turned toward me, as if she were trying to tell me she approved of that prayer. I left the room and told my wife, who could not bear to see her in such a miserable condition, about the slight movement as I prayed. That night, I dreamed I was visiting her in a prison cell. I asked her if there was anything I could do to help get her out of there. To which she replied, "No, only my husband can do that; but don't worry. I'll be out of here soon." I told my wife about this, and she merely sighed. Three days later, my friend died. I am not sure what kind of dream that was. Perhaps it was for the same reason I had the dream about my deceased wife, to console me. Perhaps it actually was a precognitive dream.

The Assassination of President Kennedy: Another dream that was definitely precognitive involved the assassination of President Kennedy. The night before the assassination, I dreamed I was in a small, dark room with several other people. In the dream, a woman stood at a tall table, and a girl, about five or six years old, was by her side. The child came over to me and started to touch me; but stopped when the woman, evidently the mother, told her not to get me dirty. That was the end of the dream.

The next day, about Noon, a co-worker came to my office exclaiming that the president had been shot; and he suggested we go to his home, which was close to where we worked, to watch the news on TV. We drove there and went into a small, darkened room where a woman, the co-worker's wife, was standing at her ironing board, pressing clothes. Their daughter, about five or six years old, was sitting in front of the TV, which was relating news about the assassination attempt. A neighbor woman was also in the room. The couple's daughter, who had been eating chocolate and had it on her hands, got up and walked over to me about to greet me, but was stopped when her mother told her, "Don't get Mr. Nagell dirty."

To me, the dream was definitely precognitive. The strange part was that in my dream there was no indication that the president had been shot. Perhaps the news of the assassination was in my dream, but it was too terrible to think of; and I refused to recall that part. Perhaps this indicates that dreams can foretell future events, but usually only as the future affects us, whether directly or indirectly.

The orange clothing: Sometimes dreams are continued. One night I dreamed I was with a group of dead people waiting for a boat to take them across the lake to their final reward or punishment. When the boatman saw me, he told me I was not dead and could not get in the boat. Suddenly my Uncle Henry, who had died several years earlier, was at my side. He took me to a place where a river entered the lake. At that point, the river was only about two feet wide, which I could easily cross. Then I was in a train station with a large group of people carrying luggage, milling around. I walked over to the boarding gate where a uniformed attendant was standing. As I entered the gate, I asked the attendant why all the people with luggage were not getting on the train. He said they could not because they would not let go of their luggage. The gate was only wide enough for a single person, without luggage, to go through. I was allowed to board the train. Then the dream ended.

The next night I got off the train and I was at the entrance to Hell. A person that seemed to be a regular human being greeted me, telling me I was supposed to get a free tour. At the end of my tour, I was getting a debriefing from the boss of the place. He asked me if I enjoyed the tour. I said it was very interesting, but I wondered why all the imps and demons were dressed in orange. I said I thought they always were dressed in red outfits. He said, "We don't like that color down here." He tilted his head back and rolled his eyes. I immediately knew what he meant. He was referring to the red of Jesus' blood.

When I woke up, I knew what the dream meant. For some time before the dream, I had been wondering about the true position of Jesus in heaven. Was he really the Son of God? The dream settled the question forever. I don't know whether this was a wish fulfillment dream, a problem-solving dream, or an actual spiritual revelation.

The morning newspaper: Many years ago, I was living in an apartment on the second floor where there were two other apartments. One other tenant received the same morning daily newspaper that I did. When I was the first neighbor going downstairs, I would get the paper for both of us and drop it in front of my neighbor's door. If he went down first, he did the same thing for me.

Three mornings in a row, the neighbor evidently got up first and brought the paper to my door. For these three days, I had dreams just before waking of what I saw on the front page of the papers. When I opened the door to the hallway, the morning paper was there; and I saw on the front pages pictures of the same things I had just dreamed about. If it had only happened once, I would not have remembered it. Since similar dreams happened to me three days in a row, I have never forgotten them. They might have been precognitive or they might have been telepathic messages from my neighbor. I could have been semi-conscious with my mind receptive to my neighbor's thoughts.

The dented fender: Another dream that clearly was precognitive in considerable detail was about an accident that occurred a few days later. In my dream, my automobile had been damaged in an accident. It was so real that when I went out to get the morning newspaper, I looked at the car and felt the fender that I dreamed was damaged. It was in good condition. However, I could not forget about the dream. On the following Saturday, I was driving to visit friends who lived about five miles away. As I started the trip, I remembered the dream, causing me to drive more carefully than usual. I resisted getting into conversations and kept my eyes glued to the road. As I turned into the road leading toward my friends' home, I had a sick feeling, but continued on the way. As I was proceeding through an intersection controlled by traffic lights, another car swerved in front of me, causing our cars to collide. The resulting damage was to the same fender I had dreamed about being damaged.

Precognitive dreams present a dilemma. If I had not dreamed about the accident, I would have been driving in my usual manner, probably a little faster. The resulting collision might have resulted in more damage, perhaps serious injury. On the other hand, if I had been driving faster, I might have been through the intersection before the other car made a turn. Precognition seems to imply that we are predestined to do things and experience certain thing. When we seem to be predicting something, we are merely in the process of "remembering" the script. However, if we are predestined, and we can see the future and be warned about harm, and *we can change the future*, we are thwarting the predestined plan for our lives and perhaps many others.

You can learn how to use dreams to your advantage. The first step is to remind yourself about any problems you want solved just before going in sleep. The next step is to remember the dreams by recalling as much as you can as soon as you wake up and before you get out of bed. Record the contents in writing or on audio equipment so it will be available later. More details come to you as the dream is recorded consciously. You might realize what the dream meant as soon as you get some details. Try to determine if the dream was merely a meaningless assortment or if there is some message your subconscious is trying to tell you. If it solved a problem, you will probably realize its significance immediately. The Bible has some comments about dreams that help us to understand what dreaming is all about.

> **Daniel 7:1-27** In the first year of Belshazzar king of Babylon Daniel had **a dream and visions of his head upon his bed: then he wrote the dream, and told the sum of the matters.** Daniel spake and said, I saw in my vision by night, and, behold, the four winds of the heaven strove upon the great sea....

> **Job 33:15,16** In a **dream, in a vision of the night**, when deep sleep falleth upon men, in slumberings upon the bed, then He openeth the ears of men and sealeth their instruction..."

Precognitive dreams can happen, but they are not always meaningful. The trouble is we usually cannot tell the difference between the irrelevant and the meaningful dreams before events happen. Therefore, probably the reason we get them occasionally is to keep us in touch with our hyperconscious mind—to remind us that there is more to life than what most of us think is real.

Many years ago, while I was in England, I happened to acquire a book that I believe was titled, "*An Experiment with Dreams.*" As I remember, the author was William Dunn, a mathematics teacher at a British University. The author conducted many experiments about dreams, using his college students as test subjects. His conclusion was that about 40% of our dreams are about past events, 40% of our dreams are of future events, and the other 20% could not be identified as anything meaningful.

THE SOUL

Soul and Spirit

The Soul is immaterial, but it controls a physical body during life of that body. (I think the soul is the same as what has been called *the hyperconscious* or the *superconscious.*) In other words, Human Beings are souls that temporarily inhabit physical bodies. Most theists believe that the souls of all humans continue to exist after physical death in a spiritual world and will be rewarded for good deeds or punished for evil deeds in some way or some place.

The soul enters a human body probably at least six months before babies are usually born. Many professing theists believe that the soul enters the body exactly at the time the baby completely leaves the body of the mother and the umbilical cord is cut. (Belief in the first concept causes people to be against abortions. The second concept allows people to tolerate abortions.) Most, if not all, Atheists and many Agnostics do not believe humans have a spiritual component.

The word "Spirit" often is used to mean the same as the word "soul," but it also refers to the force or substance that gives life to a physical being. The word "spirit" is frequently used to mean the soul of a deceased human that has made itself known to living beings. These manifestations are also called "ghosts." "Spirit" also refers to any unseen immaterial presence, such as the Holy Spirit. Usually the meaning of the word is clear by the context of statements containing the word.

The soul is made in the image of God. (Genesis 1:6, 26) The soul is immortal. (Matthew 10:28) The soul suffers punishment after death of the body because of sinning. (Ezekiel 18:20) The soul acquires knowledge of God. (Psalms 139:13, 14) The soul can be troubled. (John 12:27) When two people love one another, their souls are knit together. (1Samuel 18:1)

How to explain the relationship of the soul to the body

occurred to me when my five-year old grandson asked me, as he pointed to the casket where his grandmother, my wife of 46 years, was laid to rest forever. He said, "How can grandma be in heaven since she is there?" My answer might not have been original or theological; but it satisfied my grandson and my wife's sister who sat next to him. It took me a few seconds to find an explanation that could satisfy a five-year old child. I finally explained, "The real grandma was using that body until it was too sick for her to stay in it. Your real grandma left it, even though she wanted to stay with us; and she went to heaven."

Believers in most religions think that only the souls who accept teachings of the religion or cult they belong to will have a pleasant existence wherever their souls might be after death of their bodies. The souls of those who do not believe in the "correct" "truth" and/or fail to practice the "correct" rituals will perish or be severely punished after being judged by a spiritual being.

The soul is completely immaterial, invisible (under normal circumstances) and not tied to the physical body forever. It can be separated temporarily from the body during physical life of a body under special circumstances; and it separates permanently at termination of physical life. The soul continues to exist in the afterlife. In this case, it is similar to information within a PC that is transferred to a disk. The disk can be removed and transferred to other computers or stored for later use, just as Memories of a lifetime are then taken into the hereafter. *This means that a person retains his or her identity, with memories of life on Earth while in the spiritual realm.* The soul has its own memory storage function. This memory stage is not part of the physical body. **All** activities, thoughts, and observations generated by and participated in by a personality are stored here forever. This memory is stored in a spiritual realm, using the soul's connection with the spiritual realm, which also gives the soul an ability to obtain spiritual guidance and protection. As part of the eternal mind, experiences called paranormal or extrasensory experiences occur through contact with minds associated with other souls—whether in living bodies or in a spiritual realm.

Many people have reported personal experiences that convinced them that the soul is capable of existence without the physical body. Doctor Raymond Moody's books about *Life after Life* report the instantaneous display of an entire lifetime of memories in an instant to people who were clinically dead during surgical operations but returned to life. Some of the clinically dead people claimed to have visited relatives and friends who had died. A few of them claimed to have "seen" a place they assumed was paradise; but they were not allowed to enter that "area;" and they were "sent back" to their physical bodies. These reports, and there were many of them, supply evidence that some part of the mind continues to exist in some form after death of the physical body.

Many every day experiences demonstrate the ability of something originating in a solid body to exist outside of that body; and its effect can be measured, felt, heard, or witnessed. Whenever we watch TV, listen to the radio, or get an X-ray, we are experiencing the effects of a physical body that transmitted invisible rays that only became noticeable when intercepted by an appropriate device. This effect is caused by transmission of energy created by an electrical source. One form of energy that is noticeable by humans without a mechanical or electrical device is generated by the Sun. The effect of sunlight energy is observed as light and detected and felt as heat.

Hypnotism

The act of being hypnotized brings out abilities not evident in normal waking states. Sometimes a hypnotized subject can perform actions that are not possible to anyone not in that state. We have records of many remarkable feats of strength, endurance, mental prowess, survival during danger, and healing attributed to the effects of suggestion, either controlled by a "hypnotist" or self-induced. It used to be called "Mesmerism" after the name of its first modern student, Friedrich Anton Mesmer. (1734-1815) There was a flurry of serious interest in the subject in the last years of the eighteenth century and the first decade or so of the following century. After that, most of those interested were doing parlor tricks with it. Occasionally some physician or other medical practitioner has used it successfully; but otherwise there has been little practical use made of the amazing power of suggestion on a person's ability to perform these incredible acts.

In 1956, a doctor reported a case of a woman that underwent major surgery with only hypnosis to dull the pain. I have read of other medical applications. One was about a man who had a medical condition that normally required taping one arm to his body for several months. The patient accepted hypnosis to keep the arm in place without any physical restraint, which he did for the required length of time. Several months later, the patient was told he could move the arm, which was healed; and there were no residual effects of the long period of not being used all that time such as there would have been if it had been physically restrained.

Even though there is evidence indicating the usefulness of hypnotism, members of the scientific community reject the concept as either foolishness or because of possible misuse and the possibility of harm to the person being hypnotized. Members of the religious community reject the concept because they claim healing done without medical intervention can only be brought about by prayer and/or making pilgrimages to holy sites. Any other kind of mental or spiritual healing, according to leaders of organized religions, probably comes from Satan. The problem with both sides is that healing has occurred because of what seems to be the results of hypnotism and/or self-suggestion.

Healing has occurred after people prayed and when believers have gone to a place perceived as holy. Several sects and cults associated with Christianity preach healing as the main part of their religions. It does not seem to matter how or to whom the prayer was addressed, or what the name of the religion is that claims the holy site. So-called spiritual healing might be the result of self-hypnotism. This does not mean that spiritual healing is not spiritual—Jesus described how to pray in a manner that is similar to what happens when a person is being hypnotized.

It would be productive if communities of religious and medical professions were to get together and seriously research the subject of hypnotism and the power of prayer, without preconceived ideas about what can happen. Healing occurs in these cases, because the subject has faith in the hypnotist or that God will answer prayers; and it can happen: actually using the power of the person's mind. The Bible admonishes believers to pray "*without doubt in our hearts.*" (Mark 11:23) Hypnotism helps eliminate doubt and allows thoughts or prayers to work what seem to be miracles.

I would like to recount some amazing results of hypnotism; but I do not have any personal acquaintance with the subject. I am sure that anyone really interested in pursuing this subject further can find information about it. I found one book source that listed thirteen books with hypnotism in the title, mostly about its fun use.

EXTRA SENSORY PERCEPTION (ESP)

In order for our souls to exist outside of our bodies, for our prayers to be "heard," and for us to receive inspirations and revelations, some form of power outside the five-sensory or normal methods must be available to us. The name given to this power is *Extra Sensory Perception (ESP).* Exactly what that is or how it works is unknown at this time. We all think that if we give something a name, the thing is fully understood. We don't know what electricity is or how it performs the miracles it does, either. We all know electricity is real. According to the testimony of many, many people, ESP is real, too.

The term ESP usually refers to an ability to perceive information without use of the five physical senses. Another more inclusive term, *Paranormal Experiences,* is better when referring to all mental or physical occurrences that are not explainable by what is currently called "normal channels." Many reputable and intelligent people have reported (often reluctantly) mysterious happenings that fall under the category of paranormal experiences.

Religious leaders that believe in the power of prayer to reach their God do not know how prayers reach their intended objective. The prayers must be transmitted to God through some form of ESP. Probably what we call telepathy is the mechanism for communicating with the spiritual world. Therefore, it is important when considering how to reach our destinies to examine the concept of paranormal experiences.

Paranormal experiences usually seem to be initiated and/or controlled by the mind or spirit of the human being involved. However, sometimes the events occur spontaneously, evidently controlled by some foreign unseen forces. A large number of people do not accept these events as facts. To those that refuse to accept them, paranormal experiences threaten the reality and validity of the "real" world. They feel that if these things do happen, then what is "real?" Our definition of the world has changed many times as scientists have given us new insights. If these things do happen, they *are* the real world.

Inspirations and Revelations: Two widely separated groups of people do not accept the concept of ESP: many members of organized religions and most secularists such as agnostics, atheists, and anti-theists. To these groups, all experiences attributed to ESP are evil, fraud, hallucinations, mistaken normal occurrences, ignorance, foolishness, or stupidity.

Theistic religionists in the Western World believe that the only form of supernatural (or paranormal) activity that can occur comes from either God or the Devil. In addition, in most religions, only approved members of a religious sect, usually the leadership, are able to receive any communication from the God recognized by the sect—others get their spiritual assistance from the Devil.

On the other side are secularists: agnostics, atheists, and anti-theists. Anti-theists vehemently reject any report of paranormal experience. Most Atheists and many Agnostics also reject the concept; although they do it less vehemently. Agnostics and atheists think only what they can personally experience is true; therefore, it is impossible to communicate with God, because *they* haven't done that.

Nevertheless, many people of all degrees of spirituality and of all religions claim to have had ESP experiences. In addition, many people claim to have communicated with God, and received answers to prayer. These experiences are actually eyewitness accounts. However, it seems that eyewitness testimony is acceptable only when it agrees with a critic's preconceived opinion; and even when thousands of testimonies disagree with his or her opinion, there are never enough to convince the critic.

Telepathy: The most commonly reported form of ESP is telepathy. Telepathy is the ability of a person to receive information from other people without using any of the physical five senses. Christians that believe prayers can reach God believe in telepathy without realizing it—some form of telepathy must be the way our prayers reach God. Praying to a spiritual being demands some non-physical form of communication. I do not think there are many people in modern societies who expect their prayers to reach a divine being through their voices, the aroma of burning sacrifices, or bodily contortions. Those who believe praying affects things but do not believe in ESP claim that the ability to pray is something special given by God. ESP *is* something special. God gave it to humans so we could advance spiritually.

Clairvoyance: Clairvoyance is similar to telepathy, but the word refers to the ability to perceive information about inanimate objects without use of any other person or without using any of the five physical senses. "Guessing" more than five cards in a twenty-five card deck of ESP cards before they are turned over in many consecutive tries is evidence of clairvoyance. Another, more unusual, example is being able to read correctly parts of a specified page in an unfamiliar unopened book.

Precognition: Precognition is not exactly the same as prophecy: both words mean the ability to foretell or predict the future. However, the word prophecy usually refers to predictions received from a spiritual source. Many people, probably from the beginning of man as a thinking being, have made statements claiming to foretell what would be happening in the future. Some of them came true; but most of these statements have proven to be false. Those that came true might have been "lucky guesses;" but they might have been actual insights into the future.

The Holy Bible contains many passages that predicted events that have taken place since they were written. Critics claim that the so-called predictions were added to the scriptures after the incidents occurred. However, many prophecies in the Bible seem to be transpiring right now. One of the most remarkable prophecies that has come true is the return of the Nation of Israel. Some of the other prophecies about a tremendous war that will destroy millions of people seem to be coming true soon.

> **Matthew 24:6, 7** And ye shall hear of wars and rumours of wars: see that ye be not troubled: for all these things must come to pass, but the end is not yet. For nation shall rise against nation, and kingdom against kingdom: and there shall be famines, and pestilences, and earthquakes, in divers places.

These tragedies have occurred individually or in groups of two or three many times ever since Jesus said those words. However, at this time, it appears that never before has it seemed that a war is about to begin that will affect every nation. Already we see terrorism striking all around the world, which will eventually lead to a world war that will include all the nations and kingdoms. The prophecy about the earthquakes in divers places seems to coming true already. Almost every week the news media reports a earthquake that causes major damage and death. Does this prove prophecies can be true? Of course not, but it is a little frightening because all the rest of the Bible passages about events that are prophesized to happen about the same time as the events mentioned in Matthew also are possible, but were not possible before the twentieth century. Some of the statements in Isaiah describe things that could happen—soon.

> **Isaiah 13:9, 10** Behold, the day of the LORD cometh, cruel both with wrath and fierce anger, to lay the land desolate: and he shall destroy the sinners thereof out of it. For the stars of heaven and the constellations thereof shall not give their light: the sun shall be darkened in his going forth, and the moon shall not cause her light to shine.

> **Isaiah 24:19** The Earth is utterly broken down, the Earth is clean dissolved, the Earth is moved exceedingly. The Earth shall reel to and fro like a drunkard, and shall be removed like a cottage; and the transgression thereof shall be heavy upon it; and it shall fall, and not rise again.

A major volcano eruption could send tons and tons of poisonous gases and molten material into the air and spread around the world, causing darkness hiding the stars, the sun, and the moon for years, maybe decades afterwards. If the pole around which the Earth revolves would suddenly start to point at a different place in the heavens, it would certainly seem to be reeling like a drunkard. The following verse from Matthew repeats the prophecy about the sun and moon being darkened and adds a prophecy that the stars shall fall from heaven, which could refer to a massive shower of meteorites or it could be referring to the Earth shifting on its axis causing the stars to appear to be falling from their place in heaven.

Matthew 24:29 Immediately after the tribulation of those days shall the sun be darkened, and the moon shall not give her light, and the stars shall fall from heaven, and the powers of the heavens shall be shaken:

Several decades ago, some scientists presented the possibility that the Earth had shifted its axis about 12,000 years ago, and could do it again. That was proposed as an explanation for how the mastodons found in Siberia had been frozen quickly enough to keep them from spoiling. The entire planet does not have to tilt. Probably only the crust of the Earth would have to shift. That would cause more severe shaking than if the entire globe shifted.

Nostradamus: (1503-1566) was a French physician who wrote many short passages called Quatrains, which predicted events that might have occurred long after his death. He wrote in such a way that the predictions could only be understood after the occurrence. Even after an incident occurred that seems to be predicted in one of the quatrains, it is difficult to be certain it was a true prediction. Two of his quatrains seem to have predicted the destruction of the World Trade Center in 2001. The two prophecies discussed below are from *The Prophecies of Nostradamus,* 1973.

Century X, Quatrain 49, (page 409): Nostradamus wrote, "*the garden of the world in the road of hollow mountains will be plunged in the tank, and forced to drink poisoned water.*"

Garden of the world: World Trade Center (WTC).

In the road of hollow mountains: Skyscrapers.

Plunged in the tank: means to sink rapidly and violently, which is what the twin towers did after they had been weakened by the tremendous fire that melted steel girders.

And forced to drink poisoned water: the air and water were poisoned by the dust and fumes from the burning buildings.

Century I, Quatrain 87, (page 63): Nostradamus wrote, "*...earthshaking fire will cause tremors around new city and two great rocks will war for a long time. Arethusa will redden a new river.*"

New city: New York City.

Earthshaking fire will cause tremors around new city: the WTC was consumed by fire which caused the building to collapse, causing tremors in New York City.

Two great rocks will war for a long time: The word rocks might refer to the buildings; because they burned for a long time. It might also be referring to a long war between two nations or two ideologies, which is occurring now between the free world and militant Islam.

Arethusa will redden a new river: Arethusa has been interpreted as Ares, the god of war. Therefore, the red river is a flow of blood from the conflict between the "two rocks."

PERSONAL EXPERIENCES

My Experiences

My belief in ESP and related phenomena developed at an early age. I do not recall any time when I did not implicitly accept such things. Perhaps it is only through early experiences or teaching that anyone really accepts and believes in any kind of spiritual or paranormal evidence. I know it is difficult, if not impossible, to convince a lifelong skeptic to even consider such evidence.

Both my mother and father discussed paranormal experiences in my presence. In addition, my mother's mother told fortunes with cards and tea leaves for part of her income. Her ability to foretell the future was accepted by faithful clients who traveled many miles to ask her assistance in their lives. The only prediction of hers that I remember is that she once said that when she died, her "girls" would follow her within one year. She had other daughters; but only my mother and my aunt Nellie were called her "girls." My grandmother died in January, aunt Nellie died in May of that year and my mother died the same year in December. Maybe the subsequent deaths were coincidental, perhaps the deaths were caused by suggestion and maybe my grandmother actually could "see" into the future. None of her other children who were alive at the time of her prediction died until much later.

At any rate, I do believe there is some extra-sensory way that people can and do gain knowledge of the future and of other things. I do not claim to understand this nor do I intend to explain it. I am puzzled by it more than enlightened—more worried than relieved. What some people call paranormal experiences might someday be recognized as normal. Certainly, few knowledgeable people claim that the present state of our understanding of the laws of the Universe is final and complete. Einstein proved to us that what we knew of the world in 1900 was not final then; and many things still puzzle scientists in the year 2007.

I already discussed some of my dreams that contained what seemed to be visions of the future. In the following paragraphs are some of the puzzling events and experiences that occurred to me and to people I knew. Individually they are inadequate evidence of paranormal experiences, but taken together and considering my own feelings at the times, there is enough evidence to convince *me* that there are things and abilities not described in any of the physical sciences, nor in any of the "standard" religions. I cannot offer corroborating evidence, nor do I suggest these incidents "prove" anything. Many other documents are available that provide this kind of information. If you have experienced similar things, you will probably accept them as truth. Once a person has such an experience, usually they are more likely to believe other people could receive or experience paranormal experiences. If you have not had one, you are less likely to accept my examples as authentic even if I had corroboration. When these things occurred, I knew they were real and not imagined.

Belief in Prayer: My belief in prayer was firmly entrenched by an incident that happened when I was about five years old. My father had given me two nickels to buy ice cream cones for my sister and me. My sister asked me if she could hold her nickel—perhaps she had a premonition about what was going to happen. On the way to the store, even though I was holding the money tightly in my hand, I let the nickel fall. We both looked for the coin without success. We returned home, with me hoping my dad would give me another coin. He did not. He rather flippantly told me to pray about it. He was not very religious. I felt he was merely evading my request. Of course, I could not do anything about it. It seemed foolish to pray about it at that time so I let the matter drop.

When I was getting ready for bed that night and saying my nightly prayers it occurred to me that perhaps I should pray about the nickel. Even at five years old, the idea seemed ridiculous. I did not expect the coin to appear magically out of thin air. Anyway, I did pray that I find the coin, thinking that maybe if I returned the next day to the spot where I lost it, I would find it. In those days, long pants for men and boys had cuffs. When I removed my pants, the nickel dropped from the cuff and rolled around in a circle, returning to my feet. I am certain the coin was in my cuff all the time. Even at the time, I felt that way, too. However, as I thought about the incident, I wondered and I still do: Would I have found the coin if I had not prayed?

My Mother's Death: My mother had been suffering with cancer for many years when the doctor convinced my mother and father that an operation was necessary, hoping to remove the malignancy. My parents delayed the surgery until the Monday after Christmas, just a month after my twelfth birthday. That morning I "knew" she would never come home. This was a certainty, not just a bad feeling. I can still feel the agony of my knowledge. I did not discuss this with anyone. I tried to cheer my two sisters, trying to cheer myself. My mother did die during surgery. When my dad and four other people came up the stairs, I knew what they were going to say. I cried just as if it were a complete surprise. Foreknowledge does not prepare one for such a shock. Sure, this might merely be an example of childish fears. I was already familiar with death since my grandmother and aunt Nellie had died that year. I did not know that my grandmother had predicted my mother would follow her within the year. My dad knew of this prediction and told me about it later. Perhaps he had somehow relayed his feelings about the predicted death to me. I do not claim this was a pure and solid case of real premonition. However, this and other similar unexplainable experiences strengthened my belief in paranormal phenomenon.

My Dad's Grandmother: While my dad did not participate in organized religious activities, he believed in God and he believed in the existence of spiritual beings and their influence in our lives. An experience he had while a child firmly implanted the belief. The night his grandmother who had been living with his family died, he heard footsteps coming up the stairs to his bedroom. The steps sounded just like the ones she routinely made. During her life, she had done this every night to check on the children before she went to her own bed. My dad said he was so convinced that the sound really came from her ghost he was frightened and pulled the covers over his head. He did not discuss this with anyone at that time, so he did not know if anyone else heard the steps.

The Faulty Starter: A very puzzling thing happened to me while in the U. S. Navy when I was working on an assault boat in the yard of our repair facility. This boat was thirty-six feet long resting on a boat sled. Another Electrician's Mate and I were checking it out to determine if there were any problems and fix them if we could. I was preparing to check the starter and stood in the coxswain's station where the controls were. This was about fifteen or so feet from the engine compartment, which was separated from the troop compartment by a wall about three feet high. The other Electrician's Mate was halfway between the engine, and me facing the other side of the boat, working on a lamp socket.

When I pressed the starter switch, the starter sounded all right and I released the switch but the starter kept going—the solenoid was stuck. Just a week before, one of my friends had a similar experience. Before the battery cable could be disconnected, the entire electrical wiring system was burned out because of the high level of electrical current flowing to the starter. Replacing all that cabling was a nasty and difficult job.

The only thing I could think of was disconnecting the cable at the starter. Suddenly I was on the other side of the three-foot high partition between the troop area and the engine area; but I do not remember getting there or landing on my knees on the deck. However, there I was, working on the starter. Somehow, I had the proper wrench in my hand; and I immediately disconnected the cable. I stood up, very happy to have saved myself that miserable job and saving valuable resources. Then the other man noticed me and asked me how I had changed positions so quickly. Evidently, I had moved the fifteen feet or so and jumped over the partition without shaking the boat or making enough noise to disturb the man that I probably would have touched in moving from the front to the engine area. Not only did I move with amazing speed, agility, and silence, but I also got the cable free in an incredibly short time. Usually those connections were corroded and required considerable effort, patience, and lubricant to loosen. As hard as I tried, I could not remember how I got to the starter in time to prevent catastrophe, which could have happened within about one minute. If this sudden burst of understanding, strength, and agility was a normal physical action, why can't I do it more often?

The Forgotten Film: My son and his family had lived in the community where I lived for several years until he moved about forty miles away. I used a drug store near my home very seldom; but I did use it once in awhile. Every time I went there after my son moved away, I had the feeling I had left some film to be developed, yet I was certain I never had left any film at that store—I used another one, which I felt was better and closer to my home. The feeling was so strong that I asked my wife if she remembered me leaving film there. Finally one time when I was there, the feeling was so strong, I decided to look into the drawer of developed photos. As soon as I opened the drawer, I saw an envelope with my son's name on it. He had left film there more than a year before. I ransomed the envelope and gave it to my son. I worried that he might have requested me to get it for him, but when I asked him about that, he said he did not think so. The date on the packet was about a year earlier, about six months before he moved away. My son had never asked me to get it. The pictures were very personal; and I doubt that he would have wanted me to see them. He just forgot about it. But, how did I know about it? It could not have been telepathy: my son would not have been thinking about it. Perhaps it was some kind of clairvoyance: evidently, I sensed the presence of the film in that store.

The Bride's Night Gown: Shortly before I married Geri, my first wife, I was walking aimlessly in the downtown shopping area when I saw her through a crowd, a block away. She saw me and waved. I was delighted to see her; but this was a surprise—it was past her lunchtime and as far as I knew, she should have been at work. However, she was very glad to see me; she had been looking for clothing for our wedding and honeymoon. Unfortunately, she did not have enough money to buy the nightgown she wanted and was wondering if it would still be there when she took some money out of the bank and returned to buy it. Then I showed up. She borrowed the money from me—she later claimed she repaid me, but I do not remember that she did. We joked about that once in awhile during our forty-six years together.

The Back of the Store: A similar thing happened many years later, which makes it and the above experience seem unusual. That morning on the way to work, I discovered a problem with the car I was driving. On a sudden impulse during my lunch hour, I decided to go to the GEMCO store that sold everything from auto parts to groceries. After I purchased the part I had come for, I felt an urge to walk to the back of the store where the groceries were. When I approached the checkout lanes, I saw Geri just about to pay for a shopping cart full of groceries. She was delighted to see me and asked me to load the supplies in her car. Neither of us usually went to that store on that day of the week.

The Sentimental Box: A year and a half after Geri died, I married Betty. She had moved to California five years before we married. She sold most of her belongings before she moved from Florida. However, she had brought a few boxes of things. Living with her daughter, she did not need to take things out of the boxes. A few weeks after we were married, she brought the boxes to our home, but did not open them. One day I asked her if she wanted to empty the boxes and start using the contents. She agreed and later I started removing things while she was at the beauty shop. Just as I started unpacking, I started weeping. I had to stop and rest awhile. I finally finished emptying the boxes. When she returned, I told her of my sadness. She said she had been crying while she packed the boxes—she only moved the things with the most sentimental value. I think her feelings were still associated with the items she had packed, and they affected me the same way she had felt.

Other Experiences: My experiences were not as dramatic as those reported in many books and other places; but they convinced me they were not merely coincidences or some purely mechanical activity. Nevertheless, I, and some of my associates who discussed theses experiences with me, have had many experiences like these. The following examples might help us understand how this process actually "works."

Doctor Rhine: Dr. Rhine, the parapsychologist from Duke University, in a public speech mentioned two tests in which he was involved. He told of some experiments that he had done with schoolchildren. The children wrote their impressions when their regular teacher "sent" telepathic signals. Before these tests started, teachers wrote the names of students in order from best to worst student; and the children indicated if they liked the teacher or not. The children that were ranked highest by their teachers and liked the teacher got the best positive scores on the telepathy test. Those with the lowest grades and disliked the teachers got negative scores—they correctly wrote the answers fewer times than if they only guessed. Evidently, the minds of the teachers and the students were not in harmony or were antagonistic to one another, preventing or *resisting* messages from the teacher. The same situation might be responsible for the lower grades of those students. (I suggest that this kind of information should be part of the curriculum in every teacher's and every student's educational program. Teachers should think positively about each student and students should think positively about their teachers if they expect to acquire the best education.)

He also told about one of his college students that usually scored high but not unusually high on scientifically conducted telepathy tests in laboratories. One night this student came to visit Dr. Rhine at his home. He said he had a hot date and felt he could really perform well on a telepathy test. Even though the test could not be conducted under laboratory conditions, the test was done as objectively as possible in a private home. The student got a perfect score—twenty five "guesses" out of twenty-five tries.

These two situations indicate that a person's attitude is the most important element in any paranormal experience. I had reached this conclusion myself, many years earlier while I was in high school. During lectures, I always sent "good" thoughts to the teacher. This might have actually put me in rapport with the lecturer or it might have merely made me more receptive to the lesson. One of my college instructors was talking to several of us students about classroom lecturers. He said giving a lecture was like throwing a bucket of water into a room with empty milk bottles. Some bottles would get water inside but most would remain empty. I responded by telling him I felt I did get something from his discussions. His expression changed and he told me that he had noticed that I was different and he felt I did profit from the lesson. In spite of this remarkable ability, I did not get many high scores on my subjects, although I usually finished in the top quarter of my classes. At the end of one four-month long course, I finished eighth out of 125. Whether it worked as I thought it did or not, I am sure my attitude helped me get better grades in school.

If any readers want to explore this area further, there are many books available on this subject. Some of them are pretty good and attempt to be as scientific as possible with a subject that seems to be unpredictable and uncontrollable.

THE THINKING PROCESS

KNOWLEDGE

Knowledge is defined as having an awareness of, or being familiar with something: certain people, or some field of science, philosophy, or other endeavor. To know something means to have the perception of and the understanding of a subject, person, philosophy, or anything else.

Knowledge is not necessarily truth; but awareness of truth is knowledge. For most of Man's history many people "knew" the world was flat; but this was not the truth. There can only be one truth about any subject—what seem to be different truths are either falsehoods or only partial truths. Preconceived ideas, prejudices, and stubborn adherence to untested bits of information can prevent acquisition of true knowledge. Lack of true knowledge, along with "knowing" untruth, can cause trouble or problems in science, politics, religion, and personal lives.

Methods of Acquiring Knowledge

The ability of an individual to reach his or her destiny depends mostly on knowledge acquired by that individual. Creativity, prejudices, attitudes and to a small extent, the intelligence of an individual are influenced by the type and amount of knowledge stored in the mind. These things are acquired through experience.

Experience means all external forms of knowing. This kind of knowledge is also called "empirical," which includes conscious awareness, your personal activities, experiments, classroom lectures, books, radio, TV, anecdotes from other people, and all other sources outside yourself. Perhaps the most important source of information is the history of mankind, including traditions. History is the record of mistakes, failures, and successes. It is foolish and dangerous for new generations to ignore or dismiss the struggles of ancestors and their neighbors to make life more bearable and give freedom for everyone to proceed and succeed on his or her own ability. It is a waste of knowledge. There is no sense in repeating mistakes proven to have brought pain and death to millions of people in the past. The expression about *ignoring history forces people to repeat the same mistakes* might seem trite, but it is true. I have heard people say, "I want to learn by experience. I don't need to read a book." This is extremely shortsighted; every book is a record of the experiences of someone or some culture, even though all written material includes personal interpretation of the facts that are presented. Reading books and comprehending the information presented can be an important part of anyone's experience, especially if books by authors with different viewpoints are used. Acquiring knowledge from different sources helps to eliminate effects of prejudices, mistakes, and viewpoints of various authors.

A person's viewpoint is dependent on several conditions. Perhaps the most significant condition is the scale of observation—for example, a picture of a rock, without any other object in the picture might be assumed to be of a small or a large rock or even a mountain without some identifiable reference showing its size. For many centuries, most people thought the world was flat because they saw and felt it that way. Sailors and people living near coastlines or at high elevations with unobstructed views could see far enough they could see the Earth was curved—if they were not confused by "popular" knowledge. Experiments conducted to gain or prove knowledge usually have some goal in mind, limiting possible outcomes to what is expected. Equipment used to gather information influences the ability to gather knowledge. Devices used to gather information are only as definitive and accurate as the experimenter thinks are necessary, and limited by the technology involved in creating the devices. In addition, the desired limits of study and accuracy might be unknown or may not be available. We know that there are sounds heard by dogs that are undetected by most humans. Color-blind people do not see things the way most others see things. Some autistic people seem to have a unique ability to gather and remember information other people are incapable of seeing or understanding. If an observer is unaware of the hidden information, he or she would not even search for it. Measuring devices, if used, frequently disturb conditions in the subject or situation being measured to some degree and can modify outcomes of an experiment. Therefore, we must recognize that even the most reliable eyewitnesses, coupled with the best material evidence available might still not present accurate and complete knowledge of any subject or situation. The best we can say about the current state of knowledge in the world is that what we currently know provides mankind with a level of peace and comfort unknown to previous recorded societies.

Experience is a fairly reliable source of knowledge, depending on the source, except experiences are never complete. Even when all available experiences about an individual subject are accumulated, they are seldom complete about that subject. Some future experience may add to or alter the extent or content of previous knowledge. For example, after centuries of studying the stars with only visual instruments, we have learned that, in addition to light emanations, there are other sources of knowledge available from the skies. We know now, through use of new devices, that there are X-rays, various frequencies of radio waves, and probably many other kinds of emanations yet unknown.

Another problem with new experiences is that their interpretations are usually affected in some degree by the observer's viewpoint and frame of reference, previous experiences, training, and emotional involvement with the new incident. The courtroom rule about the "preponderance of evidence" applies to the acquisition of all experiences; but that rule is not always a complete or appropriate measure of the total truth. One should weigh, sift, and compare with all other related information that is available, while leaving ones mind open toward new and more detailed information.

All knowing is based on what is called in legal terms "evidence." In courts of law, and in other areas of life, more than one piece of evidence is usually required to confirm truth of a situation. Evidence can be personal experience, testimony of eyewitness accounts, testimony of expert witnesses, concrete objects, and/or circumstantial facts. When sufficient evidence on a subject is accumulated and accepted, it is "proven" or "known." The degree of sufficiency varies with the nature and the importance of the subject, the consequences of error, and the nature and strength of the evidence and the witnesses. When making a wrong decision involves serious consequences, such as in a trial of a murder suspect or in the case of believing about God, it is wise to lean toward the safe side of a decision. In trying a murder suspect, the consequences of error can be executing an innocent person; so there cannot be *a shadow of a doubt* concerning guilt. On the other hand, freeing a guilty person can also have disastrous repercussions. Making errors in religious decisions might have even more disastrous repercussions. In deciding to believe in and trust God, the consequences of error might result in some kind of eternal punishment.

Whether evidence is acceptable and sufficient is almost always a matter of subjective reasoning by human beings with all their frailties and prejudices. Which means that frequently, evidence is discarded that does not fit previously "proven" situations or erroneous conclusions about the evidence. Other evidence gets undeserved strength and credibility because of previous experiences, prejudice, and personal beliefs that may be incomplete or completely or partly untrue. For example, someone that dislikes a person will readily believe allegations of wrongdoing by that person; or someone who that person likes or respects might more easily be exonerated of wrongdoing.

The concept of time and its impact on an individual varies with circumstances. Time is perceived differently at different ages and under various circumstances. Our concept of time is affected by our attitudes and experiences. When engaged in pleasant activities, time "flies;" when doing an unpleasant task, time "drags." The very young think time moves very slowly, while the older one gets, the faster time seems to move.

Much of what we think is reality is not necessarily proven true. However, for our own sanity we must accept many things on faith in those who seem to be capable of knowing the truth about reality. When one begins to doubt a great many facts about what others accept as reality, mental deterioration begins.

What is real can be determined by the continuity of accumulated personal experiences and their cohesive relationships to each other, to other people, to new events, and to historical events. That is, our understanding of what we think is reality is primarily a function of the memory. The perception of reality at any given time is based on seeing, touching, other senses and emotions, and their relationships to other events and other times. Interpreting those impressions involves studying and evaluating related previous experiences. The interpretation depends on education and training, various kinds of assumptions, habits of thinking, prejudices, fears, personal goals and desires, culture, occupation, geographical position, and the historical time.

Together, these factors establish a "point of view." The ability and the willingness to accept new information as well as the ability to absorb and utilize any information depend to a large degree on the individual's personality, extent and content of previously acquired knowledge, and existing attitudes. The perception of reality and truth for any individual at any given time is therefore dependent largely on the individual, the things surrounding the individual, and the things surrounding the incident under study. Since the ability of any individual to accept and absorb truth and knowledge is dependent on the person, many people think that there are many truths—except that they frequently think their "truth" is better or truer. This weakness of humans failing to recognize that fact has caused a tremendous amount of misery and bloodshed throughout time and around the world. Conflicts involving religion or social philosophy probably all result from misconceived concepts of truth.

Interpretation of experiences involves mental activity. As you extend what you "learn" from experience, you add to your store of knowledge by comparing the new knowledge with what you think are related events or concepts. The comparisons are prone to error because of prejudice or lack of complete knowledge. Even experts are often misled by their theories, leading them to ignore or distort facts (unconsciously or consciously) to "prove" their theories. The information acquired from other people is always somewhat suspect, and should be accepted only on trust or faith in the source *and* critical interpretation of the information. News reporters and analysts have gotten into trouble because they accepted reports as facts when the reports were not properly authenticated. (The problem usually was caused by the reporter's willingness to believe something that he or she wanted to be true.)

Any information acquired from other people through books (including this one), conversations, speeches, the news media, or other sources may be only opinions, or partly opinions, and not completely factual; or the information might be based on incorrect data. The so-called facts may be deliberate fabrications. If the information is only partly factual, or incorrect, it may be impossible to separate truth from opinion or falsehood and the entire supply of data might be worthless.

Politicians and other demagogues are responsible for much misinformation when they try to make you do what they want done. During political campaigns and other attempts to impress their will on listeners, politicians and other demagogues say many things that are falsehoods, exaggerations, distortions, or merely opinions. After this erroneous information is repeated many times with authority and vigor, many people believe it and make it part of their store of permanent knowledge.

When several people observe anything, especially if it is unexpected or unusual, there may be as many different descriptions and explanations as there are people. Eyewitness testimonies from more than one source seldom agree with each other completely because each person has individual methods and criteria for interpretation of experiences. Even "experts" testifying in courtrooms often contradict one another about the meaning of physical evidence. Any experience might be mixed up with other unrelated experiences occurring at or about the same time, confusing the facts and conclusions drawn from the experience.

Reason means using your own mind to either generalize or particularize from one or more experiences and/or inspirations. Another term meaning somewhat the same as reasoning is "thinking." This is a real form of acquiring knowledge by consciously combining various forms of knowledge and/or additional sources of knowledge to synthesize new knowledge or correct existing knowledge. By combining facts about a single incident with previously acquired knowledge, a thinker can identify similarities and create a new group or category; or place a new incident in a previously generated group.

Reason is inseparable from the nature of the person who does the reasoning. However, intelligence and experience can and usually do improve the process of reasoning. Unfortunately, falsehood can be brought into the store of knowledge with improper teaching and/or emotional concerns, causing reasoning abilities to suffer.

Invention is a mental activity that seems to be unrelated to any external source of knowledge. An invention is an entirely new thought, while creative endeavors usually are more truly integrative processes, where an apparently new thought or some device or other material object usually results from previously gained knowledge combined with an immediate recognition of a need. An apparently new invention may be an improvement or an extension of an existing idea or object; but an invention does not have a recognizable source in the area or time of the invention.

Integration is the act of bringing many bits of knowledge together to form a new useful combination. Integration involves bringing bits of previously acquired knowledge together to form a new set of facts or knowledge. The knowledge acquired this way might be traceable to previous bits of knowledge with diligent searching, but usually many of the bits are so altered during the integration that the completed product may seem to be entirely new. Examples of integration are artists' presentations such as sculpture and painting, this book, and most other books, including fiction books. On the other hand, poetry might result from invention or creativity, depending on the subject matter and the manner in which the material is presented.

Inspiration means receiving conscious thoughts from what has been called the "subconscious mind," more-or-less unexpectedly. These thoughts may result from intuition or revelations.

Intuition is often considered a psychic or mystical activity; yet it really is a burst of thought based on experiences and memories, some of which might be from long before. Most intuitive inspirations result from creative processes, which are completed after a period called "incubation," a time of subconscious activity after presenting a problem or a situation to the mind. Hunches are a more immediate form of intuition, probably a result of related experiences and study accumulated over a long period of time, called to the consciousness by a sudden urgency or an unexpected opportunity. Intuitive thoughts and hunches are usually ignored by most people and only remembered when the subconscious urging has been proven correct. Intuitive thoughts are accurate sometimes, but unfortunately, many intuitive thoughts and most hunches are not correct or useful, explaining why people are reluctant to accept them.

Revelations are inspirations involving spiritual or philosophical concepts. Revelations may be prophecy, a new spiritual or philosophical insight, or a command to do something or refrain from doing something. The recipient of revelation is seldom certain of the validity of what is received; and just as with intuition and hunches, the messages are often ignored. Those who think they are "blessed" with these messages often reject them, at least at first, sometimes because of lack of faith in the source or because of conflict with previous learning or personal desires. Of course, most thoughts believed to be revelations are meaningless and often are harmful to someone, somewhere. In most organized religions, only revelations to or about recognized authorities and able to withstand long periods of investigations are accepted. Many "revelations" were not accepted by organized religions that existed at the time of the revelation. When leaders of organized religions rejected revelations because of conflict with established dogma or disruption of the relation of the leaders with their followers, those who accepted the revelations often started a new religion, sect, or cult. Revelations can come from the "conscience" which seems to be somewhat uncontrolled by the conscious mind; and it may be a fourth type of inspiration or intuition.

Conscience is a name given to the mental function that lets you know when you are doing something wrong and/or when you should be doing something you are not doing. Some people think this is a channel of communication directly from a spiritual source and is a mundane form of revelation. However, it seems more likely the conscience is a result of previous experiences. The conscience can be developed or diminished by training from others or by personal reasoning. Therefore, it may be a function of the mind that has been trained into a pattern of thinking. Then, when behavior or thoughts deviate from that pattern, this function, or memory sends signals to the conscious mind, alerting the recipient to the deviation.

Psychic Phenomena (ESP) are believed by many people to be sources of real, true information. The process of acquiring knowledge this way usually means telepathy or clairvoyance, but often involves predictions or prophecies. Many people believe they have received messages through this method. Here again, as in most inspirations, the recipient may be skeptical of the validity of the message and disregard it. Many who have received what eventually were proven accurate messages through ESP failed to report the experiences until the messages were proven by other means to have been real and correct, which generated opportunity for skeptics to reject reports of the experiences. Nevertheless, those who have received true information from this source are likely to believe people that report their experiences with ESP much more readily.

Information from or through any form of inspiration must be considered very carefully before any action is taken because of it. (This subject is discussed more thoroughly in *DESTINY, Volume 3, under the title, "Verification of Answers to Prayer."*)

INTELLIGENCE

Definition of Intelligence

Our ability to know anything depends on our intelligence. Intelligence is defined as the ability to learn, understand, and remember new and difficult situations and to understand and solve new and difficult problems. Another characteristic affected by intelligence is the ability to predict the outcome of future events based on current information. That is, once we learn a stove is hot, we are very careful when around stoves ever after. This characteristic can cause difficulties; because not all stoves are hot and it may be necessary to go near or gently touch the stove to be sure of its temperature. As defined, the word applies to all living things, possibly including vegetation and some microorganisms. We think that humans have a much higher level of intelligence than these other living things; however, one or more of them might have levels of intelligence equal with or superior to human beings.

In order to satisfy mechanistic urgings, the degree or level of intelligence has for about a century been calculated by tests that assign a numerical value called "Intelligence Quotient" (IQ) to people that have taken the tests. An IQ of 100 is normal or average, by definition. (The normal level is that of most of the people tested.) It is difficult to measure variations of more than 50 higher or lower. A person who would test 50 points or more below 100 would have difficulty with any test. Most tests do not even have the capability to test levels over 140. One important point about intelligence tests is that they only identify the lowest level possible for the test participant. Because of unique conditions: illnesses, distractions, language or reading difficulties, or many other conditions, standard tests probably do not usually indicate a person's maximum level of intelligence. Anyone graded by a test might actually be more capable than demonstrated by the results.

The level of intelligence is demonstrated by the ability to perform simultaneous thoughts and solve difficult problems quickly and effectively. An example is shown in the following anecdote. A man told me that his daughter was mentally challenged. He said that if he asked her to go to the kitchen, lock the back door, and bring him a glass of water, she would either lock the door or bring him the water, but never would she do both.

Put in a slightly different way, the number of mental tasks that can be handled or processed simultaneously demonstrates the level of intelligence in humans. In electronic systems this is called, "feedback loops." From my personal observations, I think the average number of feedback loops in humans is four or five. For example, we can talk, look around and interpret activities around us, be thinking about something we will be doing later, and swatting at a flying insect: four simultaneous mental activities.

We can see results of intelligence in many animals that are associated with humans. The intelligence of animals is inferred by observation of how they behave under conditions of stress compared with other members of the same species, and in some tests, compared to how humans would do something. It is obvious that different animals within one species have different levels of intelligence, which is accepted without too much argument by most scientists. It is also obvious that different humans, even within a single family, have different levels of intelligence, which is not so readily accepted by scientists, especially sociologists. When intelligence and different humans are discussed in the same venue, emotions and prejudices are more important than observations. The level of intelligence affects the standing of a person in society and becomes an important issue in all relationships. Therefore, any suggestion or data that implies a person or an ethnic group might not have a high IQ is politically incorrect. However, it is always permissible to call political opponents "stupid" or "incompetent."

If test scores and demonstrations of intelligence vary from individual to individual, there must be some reason. Sociologists say it is because of environment. Even where siblings appear to have the same environment, sociologists claim (evidently to support their theory) there must have been some environmental differences—either pre-natal, during birth, or during childhood.

Other experts claim the variations result from heredity and other unchangeable conditions, which stimulates a great deal of controversy by implying that gender, and ethnic and racial ancestries are important contributors to a person's level of intelligence. Members of scientific, athletic, educational, or media groups that suggested or implied that the average level of intelligence in any group of people might be lower or higher than any other group have been severely criticized and often have lost their sources of incomes. Exceptions to this statement occur within various social or political groups when the average intelligence of favored groups is claimed to be higher than less favored groups.

Even the most intelligent people make serious mistakes. I have worked with some very capable people. I believe that one of these was extremely intelligent, probably the most intelligent person I ever met. This man could solve problems in minutes that other men in his field had stumbled over for days. These other men were not dummies, just not as intelligent as the first one I mentioned. In spite of great mental capability, the man made expensive mistakes in business and tragic mistakes in his personal life. The smartest people are not capable of being correct all the time in every phase of life when decisions require expertise not within their specialties and especially when emotional matters are involved.

Factors of Intelligence

Whatever the reason, the level of intelligence seems to be an innate and stable condition for any individual, especially after reaching maturity; although there are studies that indicate it can be increased slightly at any age. Since the numerical value assigned to a person's intelligence is based on tests, any apparent increase in the number might only reflect an increased ability to take tests. One important thing to remember, as mentioned earlier, is that any test only indicates a minimum level of ability, not a maximum level. Intelligence tests depend on more than a participant's innate mental ability. Knowledge, judgment, wisdom, and creativity are also significant. All these factors depend on one another.

Knowledge: People with thorough knowledge of some subject or area might seem to be very intelligent while those with limited knowledge of that subject might seem rather stupid regardless of their real intelligence. Persons with backgrounds different from the average of those being tested could test at higher or lower levels than others with the same innate intelligence. Standard intelligence tests discover the subject's ability to store, retrieve, and utilize knowledge. Solving real, situational problems, the true test of intelligence, is much more difficult to measure objectively. This characteristic is quite apparent in politicians. During election campaigns, they are judged by their abilities in the first group: they are frequently compared with opponents who have less experience in the areas of interest at the time of the campaign and appear to be more capable. However, once elected they are judged by their abilities to solve real problems, where they might be terribly inadequate.

Memory: Knowledge is of little value if it is not accessible to the conscious mind. All knowledge acquired by a person is stored in the memory. Recalling memories is not easy; and often memories cannot be recalled when they are needed. Probably, most people can recall something *important* that happened during early childhood; but under hypnosis, people have recalled insignificant details of many events that occurred then. Experience with autistic people (those with a unique form of mental limitations) has shown that some of them have remarkable memories and can demonstrate incredible abilities to remember and transcribe information and activities, usually only in one subject or area. One autistic man was able to draw the entire scenery of Rome, Italy within a day after having flown over the city once. Some autistic people can solve mathematical problems faster than a computer expert can enter the information and receive the correct answer. One autistic man can remember any musical piece and play it on a piano repeatedly after hearing it played only once.

Judgment depends on knowledge, personal experience, and inputs from others. Intelligence can contribute to making good judgments but it seems to be of less value in making judgments than other factors. Intelligent people frequently show poor judgment because they do not always consider inputs from outside sources, especially from others who they think might not be as competent as they are. One thing that might affect a person's judgment is the failure to recognize personal conflict within or about a problem and the need to resolve that conflict. Compromise is often required for good judgment and highly intelligent people often have personal attributes that preclude compromise of any kind. In addition, intelligent people might see too far into the future, foreseeing problems not apparent to others and resist making judgments that could be incorrect in the future.

Wisdom is something like judgment, but is more complex because it appears to be innate, although it also seems that wisdom increases with age—in some people. Perhaps the apparent increase is because judgment and wisdom not only complement each other but they act synergistically and improve the usefulness of one another. Both words are used in definitions of one another. They both indicate discernment. However, good Judgment implies objectivity, while having wisdom or being wise implies having superior subjective decision-making abilities.

Creativity is useful in the practical level of intelligence because solving problems is a creative process. It would be difficult to devise an IQ test that eliminated the effect of creativity. Creativity conflicts with judgment because it depends primarily on lack of pre-judgment. Knowledge and intelligence are significant factors in creativity; because generating a creative solution for a problem demands understanding of the problem and having the personal ability, tools, or techniques necessary to solve the problem. On the other hand, being creative when taking tests can result in lower scores—a highly creative person might see solutions not apparent to the test makers, and provide answers that cause the total score to be lower than it should be.

The Value of Intelligence

Often those who are claimed to be intelligent are merely popular with the claimant. As stated before, in our society intelligence is valued highly and attributed to favorites while non-favorites are considered to be rather dumb. This is especially true about political issues. Every statement by a non-favorite is scrutinized and criticized, while even stupid sounding comments by a favorite are claimed to be brilliant observations or clever bits of repartee. Intelligence is too rare and useful an ability to be used as a social or political weapon. Society needs all the help it can get in surviving conflicts with nature and evil forces from other humans and improving the quality of life for all of us. The world needs all the intelligent people that are available, regardless of their race, gender, ethnic background, religion, politics, or other physical or social characteristics—as long as they do not have preconceived and biased ideas and are willing to accept ideas contrary to their own beliefs.

It does not matter if one group or another or one political party has the most people with superior intelligence, judgment, wisdom, and creativity. If people possess these characteristics, knowledge is not nearly as important, because people that have sufficient intelligence and are not prejudiced against the data that is required can always acquire information needed to solve the problem. (This is the concept of on-the-job training and acquiring experience.) Groups do not create anything. Individuals create. If any group has a higher or lower percentage of highly intelligent people, individuals with great intelligence in any group should be encouraged and supported in their quest for knowledge and their efforts to produce worthwhile results affecting all of us.

We must resist the efforts of socialists to make everyone the same, all with the same mediocre education and completely equal opportunities. Probably every important advancement in the welfare of society was the result of only one or a few individuals, even if implementation of the advancement might have required participation by large numbers of other people.

Intelligence tests only indicate a few mental characteristics. Some new tests have extended the number further. However, intelligence may be merely the accumulation of many talents and skills physically innate to an individual and acquired during youth and into later life. There have been studies that show, as mentioned earlier, that children who were trained in musical skills before they are twelve years old show unusual growth in certain parts of their brains. It also appears that children who had great numbers of worthwhile experiences at an early age do better in school and later in work positions and personal relations.

Forms of Intelligence

I have observed that people who achieve success in any field have a great number of skills or talents, more than just in the few areas examined in the usual intelligence tests. Howard Gardner in his book, *Frames of Mind*, (Basic Books, 1983) listed seven kinds of smart. I think there might be at least thirty of them: I listed thirteen below. The first eight are based on the above reference.

Memory: can recall many events, names, and other items.

Linguistic: is early avid talker, writes poetry, and makes up stories.

Musical: listens to, enjoys, and makes music.

Logical: good in math, categorizing, and recognizing patterns.

Spatial: visualizes, good in visual arts.

Kinesthetic: moves gracefully and handles objects skillfully.

Self-Knowledge: recognizes ones own capabilities and limitations.

Social-Knowledge: notices differences and qualities in people.

The following five skills are based on my own observations working with people in various jobs and other activities.

Time: understands significance of historical events with relations of dates and events in different parts of the world, discerns difference between transience and permanence, yesterday and day before that, today and tomorrow, knows how to schedule and work with schedules, realizes significance of eternity, and understands relation of time and money,

Extra Sensory Perception: has ability in one or more of the ESP capabilities such as: telepathy, clairvoyance, preconception, and telekinesis.

Color Sense: has excellent color memory and color distinctions, can match shades of gray with matching colors, and shades and tints of colors with shades and tints of other colors.

Foreign Languages: can identify differences in sounds of different languages, recognizes importance of physiological activities in speaking foreign languages, can reproduce foreign sounds and can speak learned languages with fluency.

Transcendental: this is a form of thinking that transcends the world of experience and into a theoretical level. Other words that have similar connotations are mystical, imaginary, conceptual, abstractive, and insightful. The word "transcendental" can apply to theological subjects, but it also applies to science, mathematics and many creative activities such as art, writing, and acting. It includes any form of thought that is not directly linked to the physical world. For example, simple arithmetic is not necessarily an area requiring transcendental thinking, but theoretical mathematics is. The other forms of intelligence are probably shared with other animals to some extent, but this form is probably unique to human beings. This may be the representation of our spiritual selves—our souls.

Reasoning

Reasoning refers to the way intelligence and knowledge are applied or used. There are seven different kinds of mental reasoning. While most people seem to use only one kind of reasoning ability, some people are capable of using more than one. Most people can acquire ability to use at least a few other kinds of reasoning with proper training. At this time, schools are more interested in developing creativity, which is fine if done correctly; but reasoning skill is more important and almost completely ignored by most teaching methods. (I think that is because teachers are afraid students might contradict them or challenge them in some area.) Different methods of teaching are required to identify and properly and completely develop the minds of people displaying each method of reasoning. Providing separate kinds of teaching is probably not possible in most schools; but education facilities should recognize the needs of different kinds of people, and try to help all students.

Politicians and demagogues of all kinds have resented and resisted efforts to teach any kind of reasoning; and they have resisted efforts to bring reasoning skills into the daily lives of people. If all people used reasoning in making all decisions, propaganda would not be able to swing opinions and decisions. Demagogues do their best to prevent their audiences from using reason. Every political campaign is a demonstration of attempts to restrict reasoning by lying, confusing the electorate, and creating fear and hate. Religious leaders, too, do not want their charges to be too critical and accept new ideas contrary to the denomination's theology. The success of propaganda depends on suppression of reasoning by members of the audience.

Various kinds of reasoning that I observed in people I was associated with are listed below.

Analytical: tests, verifies, evaluates, is critical of defects in things, people, and ideas; challenges authority, and tends to be combative and competitive. Analytical people learn by dissecting information and must have access to enough information to do this. They will challenge teachers and criticize the source and truth of information. Teachers of analytical people must be tolerant of criticism and give analytical students leeway with guidance to keep them from getting too far away from subjects being studied.

Mathematical: sequential thinker, considers things in some kind of order, likes numbers, is good in arithmetic and mathematics, and understands relations of things that are related to each other in some kind of order. Mathematical people learn things in sequence and must be taught in what they think is the proper order. History should be taught in the sequence that events occurred. These people have trouble with the whole word approach to language and with the so-called new math.

Integrative: coordinates, correlates, combines related items from various sources. Integrative people can learn by almost any method; because they can see relationships between all sorts of information not directly related to each other. They often have difficulty with multiple-choice tests because they can identify relationships not apparent to those who devised the tests; and they might select answers that seem more correct to them than the answers that are expected. This is similar to creativity; but it is more related to theoretical concepts instead of mechanistic concepts. (I think more women than men have this characteristic.)

Creative: artistic, assembles and arranges things in what is thought to be more pleasing orders, develops products from existing concepts, takes things easy, and works hard when occupied at tasks he or she considers to be pleasant and important. Other tasks are treated more cavalierly. Creative people learn by connecting information from various sources. They will create stories about the subject matter, perhaps including themselves in fantasies. (Everyone that speaks is performing a creative act because each statement requires the assembling of words in meaningful statements.) Teachers should encourage the creative approach; but they must be able to help the student minimize far-reaching destructive fantasies.

Innovative: this form of reasoning is similar to integrative and creative reasoning but is an independent form of reasoning. Innovative people arrive at conclusions not previously known.

They generate new ideas, draw from imagination, suggest new recipes and plans for many things, and tend to be iconoclastic, more inclined to fantasy, lovers not fighters. Innovative people learn by adding to existing information. They will expand on what teachers provide and sometimes show their teachers something new or never previously considered about lessons.

Practical: mechanically oriented, understands complex devices whether natural or manmade, and recognizes relationships of similar functions in different areas. Practical thinkers progress at their own paces, searching for ways to apply new information to their own lives. Sometimes they are behind the rest of the class; and sometimes they are ahead. It is seldom that they will be completely in step with academic requirements. They must digest everything they are taught and some information requires more consideration than others do for them. They learn more easily using hands-on in laboratory or shop classes. Sometimes they will find classes boring because they do not realize the value of the lessons to their lives.

Philosophical: thinks just for the sake of thinking. Philosophers are always asking why things are the way they are or why they are supposed to be done in certain ways. They are more likely to be planners than doers are. They understand theoretical concepts, and love to learn just for sake of learning. They tend to behave in what they think is a morally correct manner, whether spiritually inclined or not. Their minds will wander from what is taught and sometimes wonder why the information is considered important. Teachers must provide reasons for what is important.

CREATION

THE CREATION PROCESS

Big Bang is the name of a modern theory about the beginning of creation of the Universe about 15 billion years ago. This theory states that the Big Bang marked the beginning of space, time, the four fundamental forces, other physical laws, galaxies, stars, planets, and other items of space including living beings. This was an instantaneous "explosion" when, as the theory goes, a tremendous amount of energy in the form of subatomic particles was suddenly expelled into what we call empty space. Astrophysicists have gathered a lot of evidence supporting that theory. Stars and galaxies appear to be moving away from some place that probably was the birthplace of the big bang. So far, no other theory explains astronomical observations quite as satisfactorily. The greatest amount of astronomical evidence indicates the start of the Universe did occur at the time estimated for the big bang.

It appears that the big bang theory has credibility; and many people accept it. However, even if there was a big bang, we are not sure if it was only a sudden instantaneous condition or if it continued for some longer period. Some scientists think that some other condition or force continues to produce new material and energy. No secular scientist has come up with a theory that explains where the energy came from or continues to come from. In addition, not many theologians or other spiritually inclined people have explained from where they believe the energy originated or continues to originate, except to say, "It came from God." Both scientists and theists believe in their explanations strictly on faith.

Creation of the Universe

The big bang requires a new definition of "explosion." It could be called the "big *anti*-bang," because it was the reverse of an explosion. By definition, an explosion occurs because of a sudden release of energy being converted from material, not the conversion of energy into material.

The big bang generated particles of energy that were scattered away from the original source at an extremely high velocity. Then, these particles accumulated into larger clumps of energy, forming into atoms, which were the first "solid" objects to be created. Atoms have measurable weight, called "atomic weight." Hydrogen is the lightest atom, with an atomic weight arbitrarily set at aw1. Other atoms are heavier with atomic weights approximately multiples of the weight of hydrogen atoms. (See discussion of *The Periodic Table,* following this section.)

These atoms collided with one another, generating enough heat to fuse some of them into atoms of helium, which collided with one another and the hydrogen atoms. These combined atoms attracted one another and formed into primitive stars. These stars continued to gain material, compressing their centers into more compact atoms accompanied with tremendous heat, forcing hydrogen and helium atoms into heavier atoms (elements) with atomic weights up to carbon. The heat within the stars kept increasing and the simple first generation stars exploded, casting atoms, molecules, and complex material into space. The processes of compaction and explosion created atoms and molecules with atomic weights up to the level of iron.

The material, some pieces very large, accumulated into second-generation stars where the remaining elements were developed—some in the cores as the stars aged; and then more elements formed when the stars exploded, which sent debris from the explosions into space. The debris was composed of atoms, molecules, dust, and large objects that combined with debris from other star explosions and perhaps some of the original creation material. Third generation stars containing the rest of the elements, some surrounded by rotating planets and small particle of debris, were born. These stars were more stable.

The Solar System is composed of this kind of material, which means that the Sun and all other items within our system, including our physical bodies, are made of particles from at least two sequential star explosions.

The Solar System

As the unstable stars exploded, probably in a roughly spherical mass of fragments, their material spread into space. Debris from each star probably mixed with debris from other stars, colliding and changing the directions and speeds of the individual pieces. Debris began to concentrate into clumps that continued to travel together in the same direction, held together by the gravitational force of the growing star systems. Everything began spinning in separate eccentric orbits around the centers of various masses.

In the star group that became the Solar System, objects close to the center accumulated into a very large mass with an increasingly stronger force of gravity. The center mass became very large because at first this was where most of the debris accumulated; and then because it had such a large amount of material, its great gravitation attracted more and more material as it developed into a star. (The Sun is 332,000 times larger than the Earth.) The pressure of the material on the core caused tremendous heat to be developed and the material became a very hot and brightly shining star.

At first, all the objects were traveling around the center in separate planes with irregular paths, causing many of them to collide and form into larger independent masses. Objects that were traveling fast enough and were far enough away from the center to keep from being pulled into the center mass continued to spin around the center in separate semi-elliptical orbits. Because their orbits intersected one another occasionally, they collided with each other, forming new, larger masses with new orbits. Gradually, the objects settled into a plane that kept most of the objects from colliding. Equilibrium was reached with most of the debris settled into the Sun and planets in the same plane in roughly circular orbits around the sun, minimizing further collisions.

Comets are probably some of the original debris that made up the Solar System. They are moving in extremely eccentric orbits in planes outside of the planes of the planets. They continue to escape being attached to any of the larger bodies, probably because they are too small to collide with another object; and they are traveling too fast to remain as satellites. Eventually, all the comets will be drawn into collisions with one of the planets, or be drawn into orbits around a planet.

The Earth

As the Earth was forming, objects continued to collide with each other building into the material that makes up our planet. Because of great heat being generated from the impacts, solid materials were in a fluid state from the core to the surface because of the heat of impact and compression. The material was hot enough to keep all of it liquid, allowing the material to form into a sphere, with the heavier materials settled closer to the core. After a considerable length of time, the Earth cooled and started the chain of life that finally evolved into the humans of today. In between the major steps, there were many other relatively minor steps.

A planet with the capability to support human life has to be the result of a series of events that lead to creation of something like our Earth. The development of life of any kind, even a simple single cell, according to sophisticated experts, requires specific combinations of liquids, gases, and electrical activity. The development of human beings demanded additional unique environments and other conditions. Most of these experts say it's just a result of chance what happens and how long it takes. Many theists believe the process was the result of a plan by a Supreme Being; and it was completed into a few days.

The size of our planet, its composition of minerals and other elements, its distance from and relation to the sun, the size of the sun, the location of the moon, and other factors cannot be deviated from by much and still maintain life on Earth. Earth exists with and in a special environment. We think that the situation on Earth, or something very close to this situation, is necessary for complex life forms (as we know them) to develop.

If a planet were too small, it would cool quickly and lighter weight elements would fly off into space. Even on earth, some material, especially hydrogen and helium escape earth's gravitational pull. If something like the Van Allen belt surrounds the planet, the material returns to the planet. (This is what causes the Northern Lights.) Otherwise, with all the hydrogen gone, the planet would eventually be without water and other material necessary for life (as we know it) to form into the complex animals that are human beings. This is what probably happened to Mars, since it is smaller than the Earth and does not seem to have anything like the Van Allen belt at this time.

If our planet were too large, it would continue generating and retaining heat (as if it were a miniature sun) and be unusable by most life forms. The Earth had to be close enough to the Sun to receive enough energy for growth of plant life to support other life forms developing, but not so close as to keep the planet too hot. Earth had to be revolving around the Sun at the correct distance and rapidly enough to keep it in the same orbit indefinitely. It had to be rotating on its axis to allow energy from the Sun to be distributed more or less uniformly around the planet. It may not be vital, but having the axis of rotation slightly different from the plane of its orbit around the Sun provided a series of weather changes that make much more of the Earth's surface available for life. Some astronomers think the satellite we call Luna is an important part of our planet, acting like the tail of a kite or a sea anchor and preventing excessive wobbling of the Earth.

Other Star Systems

Many people believe creation of the Earth with its environment and the life forms that presently exist on it was an "accident," If it was an accident, it would not be likely to occur again anywhere, considering all the factors and conditions that must come together at the same time.

On the other hand, if creation were a consequence of natural forces or the work of some Supreme Being, development of planets like Earth would probably occur frequently. If another star system started, or will start, with approximately the same amount, kind, and character of debris as ours, then it is likely that there is, was, or will be a planet in that system similar to ours. Then, if development of life is a natural occurrence or the result of creation by a Supreme Being, many other planets with life, as we know it probably exist throughout the Universe. If so, some of them might have life forms that resemble us, if our forms have merit giving them unique abilities to survive and develop.

At the present time, it is estimated that there may be as many as 400 Billion stars in the Milky Way galaxy (the name of our galaxy) and there may be at least that many galaxies with just as many stars in each of them. (That means there are 16 followed by 22 zeros stars. We do not have a name for this amount; a million has only 6 zeros.) Some very courageous astronomers have estimated that one percent of these stars may have planets capable of sustaining life, as we understand it. That does not mean they do have life, merely that they could have life. That is, a great many planets in our galaxy might have intelligent life on them.

Our Sun and planetary system, which scientists believe started at least 4.5 billion years ago, (Some estimates go as high as 8 billion years ago.) is not as old as many stars in the space around us. If older stars have systems like ours, they may be older by millions of years. If the process of life is a natural and repetitive occurrence, (or if it was created by a Supreme Being) Man probably is not the only intelligent physical being in the Milky Way galaxy or in the Universe.

Other Intelligent Life

Perhaps we have intelligent neighbors somewhere close, by intra-galactic numbers. They are probably at different levels of mental, social, and technical development: either less intelligent than we, nearly the same as we, or more intelligent than we because of more ancient development or more rapid evolution.

Skeptics without humility doubt that any beings could be more intelligent than we are. Skeptics base their doubts on the fact that we have no evidence that they exist. They think intelligent life anywhere in the Universe would have tried to communicate with us. There could be many reasons why they would not. Perhaps they are not at all interested in us. Maybe we are not intelligent enough to associate with them. It might be that they are afraid of our violent nature.

The Bible contains at least one statement that might refer to life or some kind of existence on other planets. Jesus said, "*In my father's house are many mansions…*" Bible scholars usually say this means some kind of spiritual existence; but it could also mean the many mansions are other planets.

John 14:1-3 Let not your heart be troubled: ye believe in God, believe also in me. **In my Father's house are many mansions**: if it were not so, I would have told you. I go to prepare a place for you. And if I go and prepare a place for you, I will come again, and receive you unto myself; that where I am, there ye may be also.

During the last half of the twentieth century, and into the twenty-first, many people claimed they saw or experienced something that indicates beings not of Earth visited our planet.

One night I witnessed a brilliant light in the sky that made no noise and was traveling very fast. In a minute or two, an Air Force plane from a nearby air base took off in the same direction. I was interested then, and waited to see if and when the plane returned, which it did about ten minutes later. Airplanes did not normally take off from that airport after dark. At that time, the Air Force attempted to intersect unknown flying objects and identify them, so I became sufficiently interested to stay outside and waited for a while. Shortly after taking off, the planes returned to the airport. This was a UFO—an unidentified flying object, which does not necessarily mean it was an extraterrestrial object, just that it was not identified.

Another night I was in an airplane that passed closely by two unlit (unidentified) objects. The pilots refused to discuss the close miss. They had previously said they would never report seeing a UFO because they expected they would be ridiculed.

Several of my acquaintances have told me of sighting unfamiliar objects in the sky that were either flying too fast for any Earthly aircraft or maneuvering in paths that our planes are unable to duplicate. Ancient writings, carvings, and fossils seem to show that modern witnesses have antecedents throughout history and even before modern history. (Skeptics claim the ancient evidence is not interpreted correctly; and modern witnesses are confused or actually fabricating stories of their experiences.) In addition, there are structures that seem to have been constructed by techniques beyond the scope of the present state of our own sophisticated technology and definitely beyond that of any previous known human society. (The significance of this subject is discussed later under the heading, *Multiple Universes.*)

EVIDENCE OF INTELLIGENT DESIGN

Many people, including some scientists, believe that some kind of intelligence existed before creation of the Universe; and the Universe is the result of an Intelligent Design. This means that the Universe is not the result of a series of accidents, but the result of a plan, whether according to some series of natural events or accomplished by some superior supernatural being. People who find the idea of a personal God unacceptable can believe that there was some kind of superior intelligent force that for reasons unknown to us created the Universe, other objects, and eventually human beings. These people believe that after all things in our Universe were created, this force lost interest in it and either went on to create other universes and ignore the one we live in or to cease existing.

We cannot *prove* there is such a thing as *intelligent design.* All we can do is present evidence that supports the concept; and by a preponderance of *evidence*, allows us to be confident that there was, and probably is such a thing.

Physical Evidence

It is rather obvious that the Universe exists—if we accept the evidence of our physical senses. Since it exists, either it has existed forever, perhaps in some other form; or it was created through some process. It is constantly changing and decaying, so, unless new energy and material develop somehow, it will eventually die, which indicates it is not eternal and there was a starting point. That means it *was* created.

The order of things and the balance of nature indicate planning of atoms, molecules, the basic 92 elements, the Universe including Earth, and all the many different individual forms of plant and animal life. Chemicals always form together into certain compounds when mixed in certain ways. Seeds and cuttings of plants grow into the same kind of plant from which they were taken. Animals beget the same kind, based on genetic ancestry. We can predict what will happen when weather changes and other natural events progress when our knowledge of nature is complete.

Whatever the cause, or even if the Universe has always existed, it must exist in some kind of world of its own. The big bang theory says that time, space, and the particles of creation were created in less than a second with the sudden appearance of a tremendous amount of energy. In this case, the Universe exists within some other realm. If there is another realm, then the space of our Universe probably is not the same as the "space" of the realm that contains our Universe. In addition, our time is not the same "time" as in that greater realm.

Eternal Energy and Time

The Universe could not exist without some source of energy to initiate the process of creation. There must be a source of that energy. Here again, the source of this energy would have to be God or something or someone or some group of beings very similar to God.

Part of the trouble with belief in a personal God is the concept of time. God, if such a being exists, must be outside of time, as we know it. Time, as we understand it, probably did not exist before creation of the Universe. However, there must be some celestial equivalent of time in eternity. We define time as measures and sequences of changes. If the Universe did not exist at some "time" in eternity, its creation would be a change in that realm. This is a form of or perhaps the same as what we call time. Time may be the link between the sensual world and the spiritual world. Perhaps one reason for our creation was to give us opportunity to experience and learn about time.

As we experience it, time is the measure of sequential changes in some thing or some condition: creation, life, decay, and death. That is, everything that changes takes its own amount of time to proceed with the change. The planet Earth takes a year (365 days and a fraction of a day) to circle the Sun. A day is arbitrarily divided into 24 hours, each of which has exactly 60 minutes, and so on down to nanoseconds. We have identified some standards based on physical laws, such as the amount of time for sand to dribble through a specific hourglass, the space of time between swings of a pendulum, or the time for decay of radioactive material. The length of time, in the way we experience it, is not the same for specific changes to take place under all circumstances. The time that it takes for sand to dribble through an hourglass or for a pendulum to swing back and forth depends on the pull of gravity at the site of the movement. The time it takes for these changes to take place increases or decreases with the distance from the center of Earth. On another planet of a different size, the time for sand to go from top to bottom in a given hourglass would vary, depending on the size and mass of the planet. That would not actually be a change in time, merely an apparent change.

Velocity also seems to affect the length of time for changes to take place. Experiments conducted in space travel have given evidence that this is the case. As a moving body approaches the speed of light, time in relation to Earth speeds up. At speeds close to the speed of light and the faster a body moves through space, the longer a trip to another planet or to a different star system takes with *reference to the point of origin;* but that body would age much less than a similar body left at the point of origin. For example, if an astronaut travels very fast to a distant object he might age five years during the trip and when he returns to earth, he will find all of his family and friends have aged ten years. If he then takes a new and faster space ship to the same place, he might age less than five years; but when he returns to earth, his family and friends would have aged twenty years. Albert Einstein (1879-1955) developed the theory, which is accepted by most scientists, that an object traveling at the speed of light would not age at all; but it would simply go to the edge of time—or eternity.

In addition to the effect of gravity on mechanical clocks, gravitational forces also may affect the amount of time for other changes to take place. It seems that the length of time for changes to take place with respect to Earth time decreases as gravity gets stronger. Astronomers believe that the material in a black hole (with a great amount of mass and gravity) has been transported immediately to the end of time or at least some place far in the future. A black hole is not really a "hole." It appears to be an empty place in space; but there is a star in the center with a great amount of material. There is so much material that its gravity keeps attracting more and more material until it is so massive that it prevents light from escaping. The path to this condition begins with the collapse of a large amount of material falling into a single star, perhaps with the collision of two or more star systems. It could also be that the original star was very large; and it has continued attracting material from its surroundings for millions of years. Eventually the star attracted other stars that were close enough and a new black hole with its tremendous amount of gravitation was created.

The term, "time," in a commonly accepted definition, refers to a dimension just like length, breadth, and height—a fourth dimension. From what we think we know about time, it may be that time is not a dimension. The effect of time on physical objects seems more like the result of some kind of force. It does not seem likely that a linear time would change because of changes in gravity or speed. Time may be one of the fundamental forces of the Universe. If time is a force, it probably is in some manner related to gravity, either as part of gravity or acting somewhat like gravity. If it is a force, it is constantly pulling us all in the direction of eternity, just as gravity pulls objects toward the center of gravity. As far as I know, this is a new concept. I cannot expand on this idea any farther. I developed the idea in order to help me understand what and how the intelligent force works.

Intelligent Design or God?

Those who believe in intelligent design realize that the Universe is too complex and wonderful to have happened by accident, even though they do not have any idea what the term really means. Theists believe what some call intelligent design really is what they call God. Of course, Theists don't have much of an idea about what God is, either. Unfortunately, Theists waste too much time on sectarian theology and ritual, instead of trying to understand what God has done and get some idea about how the Universe was made and how it functions. I think that if we get a better understanding of the Universe and the laws that made it develop the way it has, we might get a better understanding of what our destiny is.

People throughout history have tried to explain and verify their concepts of God (or Gods) in terms appropriate to their times. Modern attempts to explain and verify the existence of God usually include arguments based on available understanding of the sciences such as physics, chemistry, biology, botany, or mathematics. Lee Strobel's book, *The Case for a Creator*, 2004 uses what we know about the natural sciences to support belief in God. *Human Destiny* by Pierre Lecomte du Nouy, 1947 presents a great amount of scientific and practical information to prove the concept of Intelligent Design without using the term *God.* (I recommend these books for anyone interested in more study on this subject.)

Formation of Space Objects

Secularists that accept the theory that the Universe was created or started at some particular time, evidently believe that the source of original energy occurred by some kind of "accident" or by "nature" on faith, which seems to me to require more faith than believing in God. When the discussion gets to the first instant of creation, secularists talk about the original energy; but they stumble at saying where the energy came from. I always expect them to say, "And in the beginning was God," but they never do.

According to the familiar equation, ***e*** equals the product of ***M*** *times* ***c*** *squared.* The letter "e" represents the amount of energy that could be obtained; and "M" represents the amount of mass (material) that is available for conversion to energy. The letter "c" stands for a very large constant number: the speed of light. Squaring it (multiplying it by itself) makes the term much larger. The equation means that a great deal of energy can result from the conversion of a relatively small amount of material.

The equation is reversible to ***M*** *equals* ***e*** *divided by* ***c*** *squared.* That means material can be "created" by the proper application or utilization of energy. It takes an awesome amount of energy to develop even a small amount of material. Therefore, an incredible amount of energy must have been available for creation of the material in the Universe, with enough energy for evolution of stars to occur, and for life to exist and continue existing on Earth.

Mental actions of human beings generate electrical energy that can be detected by an electroencephalograph. If material can be created by application of energy, this means mental action, if sufficiently powerful and properly directed, can create or affect material objects. Most of those who believe in God assume, consciously or unconsciously, God can think and generate the appropriate amount of energy. As mentioned earlier, the Universe started with an uncountable number of little bits of energy, which gathered together to form atoms, which are the beginning of what we call solid material. The little bits were attracted to each other and are held together by the four fundamental forces.

The Four Fundamental Forces

The Universe is dominated by four forces that are fundamental to its creation and to its stability. These forces act together in a delicate balancing act, with two of them acting as attractions and the other two acting as opposing forces. These forces are *Gravity, an electromagnetic force, a strong nuclear force, and a weak nuclear force.*

These fundamental forces worked together to bring these particles together and hold them together. The first and simplest stable complete atoms were hydrogen atoms. The fundamental forces continued bringing atoms together to develop into molecules and eventually into objects, which accumulated into large masses. The later stages of this phenomenon continued for billions of years; and they continue at a slower pace because the stars are denser and more stable.

Gravity is the strongest force and can act on objects that are widely separated. It is present and active throughout the Universe. Every object, no matter how small or big, exerts a gravity force on all other objects. We can feel the effect of gravity on our bodies, and see its effect when we drop something. The moon, which is close enough, exerts a gravity force on the seas, providing the changing tides. The gravitational pull of Earth is what keeps the Moon in its orbit around our planet. The Universe acts as one huge gravity field, with all objects within it acting together to form a center of gravity somewhere near its physical center, to which all objects in the Universe are attracted; but their velocity keeps them from "falling" toward this center. The "explosion" of the big bang forced the original particles away from the center; but the force of gravity pulled atoms that appeared shortly after creation together, eventually developing into large masses of material such as stars, planets, comets, and asteroids.

The electromagnetic force draws objects with opposite electrical charges together and causes objects with similar charges to repel each other. This force is much weaker than gravity and has little effect on objects that are far apart.

The strong nuclear force pulls subatomic particles together. This force does not extend beyond the atom, but is powerful enough to overcome the electromagnetic force that tends to push similarly charged particles apart.

The weaker nuclear force destroys some of the subatomic particles.

These four forces were probably created as the first step in creation. Since these forces act through space, gases, and solid objects, they behave like a spiritual force and may indeed be physical representations of the spirit of God.

Additional laws developed at the time of creation and they continue to control the actions and conditions of objects in the Universe; and they might even have an effect on things outside of our Universe. The laws that limit the use of energy, *thermodynamics*, also limit the life of the Universe.

Energy

The utilization of energy involves the flow of energy from one body or area to another. Energy can be converted from one form into another form, or transferred from one system to another. For example, a heat engine, such as an internal combustion engine, changes chemical energy stored in fuel into heat energy. It then converts the heat energy into mechanical energy that could generate electric energy, power a vehicle, or do other work. Even though the form of energy has changed, the total amount of energy always remains the same, including the amount lost in friction and conversion processes. A nuclear explosion represents a different type of conversion: creating energy from material. (As discussed earlier, energy and material are related to each other as shown by the equation ***e*** *equals* ***M*** *times* **c** *times* **c.**) Even though it might appear that energy is created during an atomic reaction, it is merely the conversion of matter into energy representing the return to the original form of energy provided during creation of the Universe.

Thermodynamics

Thermodynamics is the study of various forms of energy, such as heat and electricity, and of the conversion of energy from one form into another. Thermodynamics is based on two laws (principles). The principles seem to be unchangeable throughout time and distance.

The First Law states that the total energy in any system is constant. In other words, energy in a closed system cannot be created or destroyed. (A system may be anything from a simple object such as a kettle of water over a gas flame, to an internal combustion engine driving an electric power generator, or up to the entire Universe.) A closed system is one that is not affected by outside forces, a condition that is impossible in our lives, because energy can be lost or gained by contact with air, other gases, or other objects or through radiation to other objects. However, the Universe is a closed system, unless there is another realm of existence that draws energy from or adds energy to the Universe.

The Second Law of thermodynamics deals with the inevitable result of energy processes. According to this law, heat will flow from a hot object only to a colder object; electricity flows from an energized source to an un-energized or a less energized location, and or other forms of energy move from one with high energy levels to one with lower levels. For example, heat energy transfers cannot continue indefinitely because the original heat will only flow to a colder object through conduction, convection, or radiation. When all objects are at the same temperature, heat transfer stops. The second law implies that any heat transfer or conversion of energy can never be completely efficient—as soon as the bodies are at the same energy level, the exchange stops. For example, if at the start of an experiment, the temperature of one object is 120 degrees and another is only 50 degrees, heat will flow from the hotter object to the cooler object; but the temperature of the cooler object might only increase to 70 degrees, depending on the relative amounts of material involved. The object that was warmest at the start will decrease to the same temperature as the previously cooler object.

In all energy transfers, exchanges, or conversions, some energy is lost in friction and/or dissipation to surrounding objects. Scientists made this more complicated by using a term called entropy, which refers to the amount of disorder or randomness in a system. For example, mixing sugar and dry cocoa together develops a random distribution of both items, and the mixture will never separate into sugar and cocoa without some special conversion process bringing the mixture back to the separate substances. The mixture represents a disordered system and the previous separated condition represents ordered systems. With energy exchanges, the loss of energy due to the conversion process represents a disordered system.

Based on our present state of knowledge, the total amount of energy in the Universe is limited; and eventually all of that energy will be lost, dissipated in space, with all objects at the same incredibly cold temperature, probably just a little above absolute zero. The temperature could cool further if there were some other cooler heat sink surrounding or near the Universe. The energy cannot be regained without some yet undiscovered activity.

The laws listed above seem to be universal and eternal. So are the following laws that complement the laws of thermodynamics:

Bodies at rest remain at rest unless acted on by an outside force. Nothing starts moving by itself. For example, an automobile will not move without having the engine on with the gears properly engaged or something, or some one pulls it or gives it a push. This law parallels the first law of thermodynamics. Regarding creation of the Universe, the law means that the big bang could not start all by itself: something or some being gave the conversion of energy into material a "push."

Bodies in motion remain in motion unless acted on by an outside force. Nothing stops moving by itself. For example, an automobile in motion will continue moving until something stops it. Putting the brakes on slows and stops the wheels from turning and friction of the tires on a surface stops the vehicle. (Friction converts mechanical energy into heat energy.) Even if nothing else is done to stop the automobile, friction within the moving parts of the wheels combined with friction of the tires against the road will eventually stop the car unless it is on a hill. Of course, hitting a solid stationary object will also stop the motion abruptly. This law parallels the second law of thermodynamics. Regarding the Universe, the objects in space that seem to be moving at tremendous speeds will continue to move in the same directions unless something or some being stops movement. The something could be the combined gravity of the Universe acting at its center.

The Quantum Theory

The Quantum Theory states that during measurements and experiments of sub-atomic particles, their movements and/or positions can be measured in more than one way, with results changed by the method of measurement—the results are indeterminate. Some physicists believe these particles are affected by the mind of the person making the experimenter. *The "mind" of God could have supplied the energy that created the basic particles of the elements and affected the way they behaved.*

Other Basic Laws

The Periodic Table: Atoms do not all weigh the same. Their weights are compared to the weight of hydrogen, which is arbitrarily set as atomic weight (aw) 1.000000. (However, as the ability of scientists to be more accurate in more detail, the weight of hydrogen was discovered to be a little more—approximately aw1.0080.)

When the names of chemical elements are listed in the order of increasing atomic weights, similar physical and chemical properties appear at regular intervals. The elements can then be arranged in a chart with 7 horizontal rows and 18 vertical columns, something like a calendar arranged to illustrate the relationships.

With the elements presented in a chart as described above in a natural sequence of their atomic weights, significant relationships become apparent. Lighter metals: sodium aw23, magnesium aw24.3, and aluminum aw27, with progressively greater weights, are on the same horizontal row. Ductile metals: copper aw63.5, silver aw108, and gold aw197 are in the same vertical column. Notice that the atomic weights of the ductile metals increase approximately by the weight of copper as the weights gets heavier. The relation of most other elements also becomes apparent when presented in the periodic table.

Creation of the Universe occurred in a sequence of events that with a few exceptions followed the order of elements shown in the periodic table. Hydrogen as the lightest element was the first element created, and other elements seem to have been fashioned from accumulations and manipulation of hydrogen atoms to form heavier elements. Helium is the next heaviest element. As the Universe matured, heavier and heavier elements were created until the ninety-two stable elements were created, like using building blocks. This indicates an orderly plan for creation, designed by some intelligence. This is the way God created the Universe, one-step at a time over billions of years.

Electromotive Series: Each element has an individual value of electrical potential. Elements can be arranged in the order of the numerical value of their electrical potentials. The electrical difference between chemicals is the basis of storage batteries. For example: carbon has a potential of +3.02 and zinc has a potential of +0.76 with a difference of 2.26. Placing these elements electrically connected on one end with the other ends separated from each other in an electrolyte (a nonmetallic electric conductor) an electrical current with a voltage of about 1.5 volts is produced. This is the formula for the common dry cell battery.

Frequency Spectrum: All energy sources, such as radio signals, light, heat, and sound, distribute their energy in regular pulses or waves, with various frequencies for each source. The pulses of each source increase in relation with each other in an orderly progression. The response of our senses to these pulses depends on the frequency of the pulses. We "hear" pulses at the lower end. Sound waves that most humans can hear occur between about 100 and 10,000 cycles per second (CPS). Radio waves occur in the range of 1,000,000 to 10,000,000 CPS. Visible light waves occur in the vicinity of 100,000,000,000,000 CPS. We detect the lower frequency light waves as blue color; and we detect the higher frequency as red color. We detect frequencies between blue and red as other colors. Humans cannot detect the other frequencies without special mechanical or electronic equipment. The highest frequencies that we are aware of are cosmic rays.

Other Considerations

The order of things and the balances of nature strongly indicate planning of the Universe and all the different forms of animal and plant life that are present on Earth. Secularists attempt to prove that order can come out of chaos by some natural process independently of any intelligent design. (Disorder cannot "accidentally" become order, as discussed earlier.) They ask us to suspend "common sense" and depend on obscure "scientific" theories or complex mathematical models that have no relationship with real life situations. If events can occur or characteristics of objects change by accident or some natural law, we can no longer depend on anything remaining the same as it has been.

Creation of life: Secularists believe life just happened to develop either by accident or by some kind of natural process. So far, it has not been shown how creation by either method could happen without some kind of intelligent pattern, although scientists are currently attempting to create a simple form of life. (If scientists actually do create life, that does not disprove the concept of Intelligent Design; because the scientists would be operating with *intelligence.*) The generation of life requires (at least) a great number of special materials and unique environmental conditions, all occurring in a specific sequence and in specific ratios. It is inconceivable that all these situations could occur without some plan. Even if all the proper conditions did exist, and life could develop accidentally, it is still difficult to believe the existence of these conditions and eventual generation and growth of life was not planned by intelligence.

Persistence of life under so many different conditions and environments indicates a need for life to exist. At every level of the oceans, deep inside darkened caves, in the tropics, at frigid poles, and other unlikely places, some kind of life form lives and reproduces. This is either "natural," or some superior intelligence has directed that life must continue to exist.

Complexities of living forms: Most of the higher forms of life such as mammals, amphibians, and most fish have the same kinds of parts arranged in similar ways, with similar functions. The structures, relationships, and functions of various parts of animals, such as the eyes, ears, heart, skeleton, and the functions of endocrine organs are so well-balanced that it stretches the imagination to come up with some "accidental" explanation. Even if we accept the secularist's concept of evolution, this does not eliminate the possibility that some form of intelligence interfered with normal reproduction of species to cause sudden changes in the bodies of animals that were "evolving." In other words, this intelligence actually caused genetic changes that resulted in birth and survival of animals that were closer to the ideal form for our souls. Earlier forms were constructed so the evolutionary changes would continue in the right direction.

Instinctive behavior seems to be driven or guided by some force or some superior intelligence not physically or mentally part of purely physical animal life; and seems to be external to the being that is acting instinctively. We do not know what instinct really is, or where in the brain it exists, if it does exist there. Evidently, it is part of the mind and is hereditary. Humans are not controlled to the same extent by instincts as the less developed animals. Psychologists and anthropologists tell us humans are not controlled by instinct, because we have free will and the ability to adjust to changing conditions.

Even if an old, inherited intelligence is built into the brains of animals at birth, the source of the ability to perform activities never taught or demonstrated to animals is mystifying. For example, a friend told me about a strange experience with his beagle dog that had lived in a metropolitan area since birth with no contact with other beagles after it was weaned. My friend told me this dog had never seen rabbits or any other similar small wild animals. One day it suddenly burst into the baying of beagles on the trail of prey. Then my friend noticed a wild rabbit running down the street where the dog could see it. This rabbit was way out of its normal area and seemed to be in panic. Rabbits had never been in that neighborhood before; and the dog had never participated in a hunt. Somehow, that dog "knew" he was supposed to make that particular baying sound.

The conscience of mankind and some other animals seems to originate outside the physical or mental nature. The propensity to do wrong and regret it, at least in most people, tends to prove the existence of what we call "conscience." It seems to be similar in all "higher level" beings. Nearly everyone with a pet animal has observed that the pet knows when it has committed a breach of acceptable behavior. Having a conscience presupposes possession of the ability to know right from wrong and to know what is right. Wolves in the Portland, Oregon zoo demonstrated an ability to commit premeditated murder, with other wolves recognizing the nature of the deed and immediately executing the perpetrator. This characteristic indicates there is some source of such knowledge and the ability to recognize, regret, and punish wrongdoing.

The human mind: The power of reasoning that exists in what we are pleased to think is a high degree in mankind and in lesser degrees in what are called the dumb animals implies that somehow we are copied from a vastly superior pattern. Our ability to use our minds seems to be an infinitely smaller form of the power attributed to God.

Accidents and Probabilities

Almost every culture in the world throughout history has had some explanation for the existence of Earth and the rest of the Universe, and development of human beings. Almost all of them give credit for creation of this world to some kind of spiritual power. Today, it's only the most sophisticated people who believe the Universe and the existence of human life was the result of "accidents." If one believes the entire process was accidental (whatever that means), then it is necessary to have *faith* in a multitude of accidents. There also had to be a lot of energy just sitting, waiting for an accident to turn the energy into matter.

Something has to cause an accident. Every action has a reaction. Every effect is the result of some action. Those that believe in the accident theory should explain what that "something" was. However, even the "first" cause must have been caused by something else—according to our concept of time. If there was and is a Supreme Being, how do we explain that? Where did he, she, or it come from? Was he, she, or it alone before creation of our Universe? With our limited intelligence and knowledge, we are unable to explain what the original cause was. A person has to have *faith* in whatever name is given to the first cause, no matter which explanation a person chooses to accept.

Secularists say, "Given enough time, anything can happen." Some of them generated numbers that "prove" there was enough time for evolution to occur "accidentally." This is a silly statement! There is one compelling argument against the present condition of the Universe being caused or influenced by the "laws of probability." Regardless of how much time is allowed, nothing can happen without some kind of cause, precedent, pattern, or possibility of occurring. For example, the chance of getting either a "head" or a "tail" when spinning a coin is one out of two. The odds go up tremendously for getting more than two or three of either one in a row. It is true that given enough time, it is conceivable that there might be a series of twenty "heads" in twenty consecutive spins. However, if someone does not toss or spin a coin, the odds are meaningless because the sequence could not develop without the first spin. The odds against winning a lottery might be six million to one; but these odds are only applicable to a person who has a lottery ticket. The odds do not mean anything to a person who does not buy or otherwise acquire a ticket. That person has absolutely no chance of winning.

The same thing applies to creation of the Universe and development of human beings. The odds do not mean anything if there is no previously existing possibility of creation occurring. The only way a possibility could exist is if there were some kind of being (or beings) or undiscovered natural condition in another superior realm that brought about creation of the Universe.

As discussed earlier, if an action could be probable, or even possible, it must have a precedent or a possibility of occurring. Where many "chance" occurrences are sequential and dependent on one another, the odds increase many times. Our present concern is the existence of human beings. If there were, in some manner, a possibility that life could start "accidentally," there is no way we can calculate the odds against creation of the first life form and then on to the development of human beings. Nobody knows the entire number of steps and their sequence required for the present condition of the Universe and the development of life on earth, especially the evolution of human beings. I think it is safe to say there would be very many steps.

Summary

Together, these physical laws are tough hurdles to overcome for the Secularists' concept of creation. The law that objects cannot start moving by themselves indicates there must have been some initial cause of creation. If it was a "Big Bang," what caused the explosion? We are certain that the explosion involved conversion of energy into matter. A great deal of energy (an infinite amount) was required for creation of the entire Universe.

If creation was caused by a big bang, the objects being flung into space will continue moving out into the farthest distances of space, spreading the stars infinitely far apart—unless some force eventually stops them or pulls them back together. This force could be the total amount of gravity exerted by the total mass of galaxies, and other space objects in space eventually attracting all other objects toward the center of gravity. This is the same as throwing a ball into the air: it immediately starts slowing down; but it rises higher and higher until eventually it is moving slowly enough for Earth's gravity to pull it back down. If this is true, the first objects to be "flung" into space might already be slowing down or even beginning to return to the Universe's center of gravity. This is like a stream of water from a fountain or a garden hose aimed to the sky that only goes so far and then stops, with the highest particles of water falling back to the ground.

Astrophysicists are not sure if there is enough material in the Universe to provide a gravitational force strong enough to pull all the material back to the center of gravity. If there is not enough material in the Universe to pull everything back together, the stars and their satellites will grow farther and farther apart and everything will eventually be at a very low temperature because all the energy will be spread around uniformly.

This means that the life of the Universe is limited to some finite period of time. Either the Universe will continue to spread apart and die, or it will be drawn back together and be destroyed in a tremendous implosion. Perhaps this is God's way of bringing all things back to Him.

EVOLUTION OR CREATION?

Evolutionism or Creationism: The word *evolutionism* refers to the concept that the Universe was created or formed and developed into its present condition during many billion years *on our time scale;* and humans are the result of millions of years of being changed from something like a modern rodent. Evolutionists believe the entire development of the Universe and, eventually, human beings was the result of a great many sequential, *accidental* "improvements." The word *creationism* refers to the concept that all of creation took place in six calendar days *on our time scale*, and human beings were created in the same form as modern humans during one of those days. Creationists believe that humans were created in our present form and mental condition by God about 6000 years ago—on our present time scale.

Even though there seems to be irreconcilable differences between the two concepts, the belief in the secular *sequence* of evolution is not incompatible with the *sequence* of creation described in the Book of Genesis. However, the time scale used by many Christians, based on calculations of dates in the Bible, is incompatible with any secular concept of evolution and the facts presented by archeological evidence. Nevertheless, Moses, who Jews and Christians believe wrote the first book of the Bible, did know this sequence. How did he know? Was it the result of shrewd reasoning based on his personal observations? Was it merely a coincidence? Did he report something he learned from some other human source? Or, did he actually receive this information from a Supreme Being called God or Jehovah?

The word "evolution" has two widely used meanings that seem to conflict with each other. One definition refers to the observable fact that occasionally life forms are born with reproducible characteristics different from preceding generations. Almost everyone accepts this fact. For example, some bacteria develop immunity to antibiotics, hybrid vegetables produce greater yields, and many varieties of canines, cattle, and plants have developed from the original forms. It is obvious that genetic changes can occur, causing "evolution" of these things; but these changes have occurred with selective breeding within individual species, either by nature or by man. Almost everyone, including Christians, accepts this definition. However, many Christians do not believe in the definition that means the ancestors of humans, other animals, and plants periodically, during millions of centuries, developed into new species of animals, either accidentally or according to some plan.

I think it is foolish to be adamant in whichever position one takes. Scientists are not scientific when they say creation occurred because of an "accident." They know nothing happens without some cause—or the result of many sequential causes, even though the cause(s) might be unknown. Right now, they do not know how the Universe and life first started. Nobody knows! Just because the various secular experts have given names to their favorite first causes does not mean they really know. On the other hand, Creationists refuse to accept the physical archeological and paleontological evidence at face value. Both extremes take their positions strictly on faith.

Many who accept the theory of evolution bitterly oppose any reference to or discussions of creationism or even of the mention that some spiritual being was involved in the process in any manner. Many Christians reject the concept of evolution; because secularists have (erroneously) used it as proof that there is no God. (It is impossible to *prove* that there is no God.) Both sides should look at the physical evidence and compare it with the Holy Bible to see if there is any similarity.

Creation and the Bible

Theists, especially those that accept the Bible, do not disagree with secularists too much on the sequence of creation and evolution. Where Theists part from secularists is in the cause of evolution and amount of time involved. I do not think the element of time should even enter into the concern of religionists. (Our time scale is not the same as God's scale.) I know that what I am about to discuss is not acceptable to many Christians; but I cannot believe that the entire Universe including human beings was created in six of our calendar days. I think this is presumptuous: we cannot define God's time by our standards. I think the original explanation for creation by the ancient Semites merely stated the sequence of events during the creation cycle, probably in pictorial form. As human knowledge increased and more sophisticated forms of communication were developed, the pictorial steps were *arbitrarily* defined as *days* in our time scale. According to the Bible, no human being was around until the sixth day; and after that, God rested; so, how do we actually know what term should be used where the Bible says "day?" Evidently, the original recorder of the incident must have had some spiritual revelation and recorded it in some ancient form of representing some language, probably on a stone or clay tablet. The term "day" must have been translated from that ancient record. The original term may have meant something else—probably the individual periods of creation with the entire sequence inscribed on seven tablets. It is likely that each tablet represented the activities that Christians believe took place in one day, but really took place over many thousands of centuries. Therefore, to ignore physical evidence about events and the time involved demonstrates a peculiarly unholy stubbornness.

As mentioned earlier, we measure time by sequential steps of certain events. One of those events is the geological changes in the appearance of the Earth. If a rock has certain characteristics according to the way we measure time, the rock must have gone through sequential changes, making it appear as if it were a certain age—perhaps millions or billions of years. Therefore, if God created the rock only a few thousand years ago by our time, it was created in a condition that appears to be much older than that. If the Earth were only six thousand or even ten thousand years old by the Gregorian calendar, it would be in a gaseous and poisonous condition; and life could not exist, as we know it. Therefore, if the world is actually only somewhere close to ten thousand years old, it was created in the condition we now know, making it seem to be billions of years old. (If that is what happened, God would have some reason for fooling us this way.)

In addition, if mankind was created on the sixth "day," the entire apparent sequence of evolution was invented by God and buried in the layers of rock also created at that time. We are expected to believe the world is real; but, according to the strictest literal interpretation of the Bible taught by Creationists, we are told to ignore the evidence.

If all the material things and beings in the Universe were

created less than 10,000 years ago in the condition they are now, we should also wonder about the entire history of the world up to the present time. If the entire Universe was created only six, ten, or twenty thousand years ago, confidence in the reality of everything is shaken. Perhaps the Universe was created at the time YOU or I, were born. Perhaps we were created only a few days ago in the condition of our present age with artificially planted "memories." Perhaps what we know through our senses is all merely an illusion created by...what? Perhaps the entire world is not real! That path can lead to serious confusion.

The geological evidence shows us that life originated as a simple form(s) and evolved to the state we see now. I believe that there is a being that we call "God;" and He created the entire Universe, including humans. I believe God has continued to work on us, evidently hoping to improve us. (We don't seem to be getting much better, considering all the hate, anger, selfishness, greed, pride, and envy demonstrated all over the world.) I choose to believe that God wants us to understand the process of creation and evolution. Perhaps by understanding the concept of creation and evolution a little better, we might actually get better as God seems to want us to be. I have enough faith to accept whatever manner God chose to bring me and the rest of the Universe into being and develop us to our present condition.

I think that it is more spiritual and reverent to believe God created the Universe billions of years ago, as it appears to have been done. I also believe that the same God is still interested in humans and our spiritual welfare.

CREATION IN THE BIBLE

The first verses in Genesis describe the creation of the Solar System and the creation and development of Earth before it was formed into its present condition. The Bible does not mention the rest of the Universe. (This omission might be important; but I can only guess what it means. It might mean that the beings with or part of God mentioned in Genesis 1:26 are only interested in or involved with our Star System and Earth.) The following Bible passages and scientific explanations for the creation of all things are compared with each other to demonstrate their similarity. People who wrote the Bible did not use the technical language of today. Even if the correct words had been used in the first version of the Bible, the translators could only use words that meant something to them. However, I think the words from the *King James Version* of the Bible convey the appropriate message. (Some terms from the *Living Bible* are also used to clarify the King James Version.)

The First Day

Bible

Genesis 1:1-2 In the beginning God created the heaven and the earth. And the **Earth was without form, and void; and darkness was upon the face of the deep**. And the Spirit of God moved upon the **face of the waters.**

Science

The Earth had not yet formed into the planet it is now; it was in a molten condition. In the last sentence above, the term "*face of the waters* from the KJV is interpreted "*dark vapors*" in the Living Bible. This term probably is more like what was originally intended; because at first, hot gases mixed with dust surrounded the earth and the Sun could not be seen. The dust and debris that were beginning to gather into what became the Earth were part of the incoherent material in the whirling mass that was becoming the Solar System.

Bible

Genesis 1:3 And God said, Let there be light: and **there was light**.

Science

The sun formed into a gigantic sphere; and it gradually compacted into its present size until it became hot enough to generate light. About the same time, the atmosphere of the Earth became clear enough for light to reach the surface.

Bible

Genesis 1:4-5 And God saw the light, that it was good: and God divided the light from the darkness. And God called the light Day, and the darkness he called Night. And the evening and the morning were the first day.

Science

The Earth was rotating, causing the cycle of day and night.

The Second Day

Bible

Genesis 1:6-8 And God said, Let there be a firmament in the midst of the waters, and let it divide the *waters* from the waters. And God made the firmament, and divided the waters which were under the firmament from the waters which were above the firmament: and it was so. And God called the firmament Heaven. And the evening and the morning were the second day.

Science

Here again, the Living Bible uses the term "vapor" instead of "water," as in *And God said, "Let the vapors separate to form the sky above and the oceans below."* This is what the secularists think happened. The scientific explanation is that as the gaseous material cooled, most of it condensed into liquids.

The Third Day

Bible

Genesis 1:9-13 And God said, Let the waters under the heaven be gathered together unto one place, and let the dry land appear; and it was so. And God called the dry land Earth; and the gathering together of the waters called he Seas: and God saw that it was good.

And God said, Let the Earth bring forth grass, the herb yielding seed, and the fruit tree yielding fruit after his kind, whose seed is in itself, upon the earth: and it was so. And the Earth brought forth grass, and herb yielding seed after his kind, and the tree yielding fruit, whose seed was in itself, after his kind: and God saw that it was good. And the evening and the morning were the third day.

Science

In the early stages of creation of the earth, hot liquid materials covered the entire surface of the planet. The Earth cooled and solidified from the surface down toward the molten center. Volcanic action commenced, bringing molten material above the surface of the liquid where it cooled into solid material.

Secular evolutionists agree that vegetation appeared before the appearance of any kind of animals. (The word grass may actually refer to molds, which came first, then came "*the herb yielding seed, and the fruit tree yielding fruit after his kind, whose seed is in itself, upon the earth.*" Vegetation converted carbon dioxide, which remained in the gas surrounding the Earth, into oxygen, preparing the atmosphere for animal life.

The Fourth Day

Bible

Genesis 1:14-19 And God said, Let there be lights in the firmament of the heaven to divide the day from the night; and let them be for signs, and for seasons, and for days, and years: And let them be for lights in the firmament of the heaven to give light upon the earth: and it was so. And God made **two great lights; the greater light to rule the day, and the lesser light to rule the night**: he made the stars also. And God set them in the firmament of the heaven to give light upon the earth, And to rule over the day and over the night, and to divide the light from the darkness: and God saw that it was good. And the evening and the morning were the fourth day.

Science

The sky became clear and objects in space could be seen from the surface of the Earth. The sun was the "*greater light*" and the moon was the "*lesser light*" after the sun began to make it shine from reflected sunshine. Other objects within the Solar System also started to shine from reflected sunshine.

The Fifth Day

Bible

Genesis 1:20-23 And God said, Let the waters bring forth abundantly the **moving creature** that hath life, and fowl that may fly above the

Earth in the open firmament of heaven. And God created great whales, and every living creature that moveth, which the waters brought forth abundantly, after their kind, and every winged fowl after his kind: and God saw that it was good. And God blessed them, saying, Be fruitful, and multiply, and fill the waters in the seas, and let fowl multiply in the earth. And the evening and the morning were the fifth day.

Science

Life in the form of "*moving creature(s)*" started in the watery liquids, presumably as simple cells and progressed from there. These cells were capable of growing, and they had the magical ability to reproduce. Reproductive capability is not essential for persistence of life; but it is necessary for the process of evolution to occur. Life, as we define it, involves three major steps: initiation of life form, growth, and reproduction, before the evolution process can start.

Including the fowl here is more difficult to compare with the secular concept of evolution. Most theories propose that birds evolved from some kind of dinosaur. Perhaps the Bible is correct and fowl were developed first and dinosaurs evolved from them. Perhaps the term "*great whales*" actually means what we call dinosaurs.

The Sixth Day

Bible

Genesis 1:24-31 And God said, Let the Earth bring forth the living creature after his kind, cattle, and creeping thing, and beast of the Earth after his kind: and it was so. And God made the beast of the Earth after his kind, and cattle after their kind, and every thing that creepeth upon the Earth after his kind: and God saw that it was good.

And God said, Let us make man in our image, after our likeness: and let them have dominion over the fish of the sea, and over the fowl of the air, and over the cattle, and over all the earth, and over every creeping thing that creepeth upon the earth. So God created man in his own image, in the image of God created he him; male and female created he them. And God blessed them, and God said unto them, Be fruitful, and multiply, and replenish the earth, and subdue it: and have dominion over the fish of the sea, and over the fowl of the air, and over every living thing that moveth upon the earth. And God said, Behold, I have given you every herb bearing seed, which is upon the face of all the earth, and every tree, in the which is the fruit of a tree yielding seed; to you it shall be for meat. And to every beast of the earth, and to every fowl of the air, and to every thing that creepeth upon the earth, wherein there is life, I have given every green herb for meat: and it was so. And God saw every thing that he had made, and, behold, it was very good. And the evening and the morning were the sixth day.

Science

Cattle and by inference, all mammals, were created after the water dwelling creatures. Creation of human beings was the latest step in the development of our planet, and according to both the Bible and science, the last step, at least, so far.

The Seventh Day

Bible

Genesis 2:1-3 Thus the heavens and the Earth were finished, and all the host of them. And on the seventh day God ended his work which he had made; and he rested on the seventh day from all his work which he had made. And God blessed the seventh day, and sanctified it: because that in it he had rested from all his work which God created and made.

Author's comments

Chapter 2 of Genesis seems to indicate that Adam and Eve were created either after or during the seventh day. It is not clear if Adam and Eve were part of the creation mentioned in Genesis 1:26-31 or if this was a separate creation. If it was a separate creation, then God had ceased resting and returned to the task of creating. On the other hand, what is mentioned in Chapter 2 might be referring to the development of modern humans.

The development of entirely new types of animals seems to have stopped a long time ago. Modern man, presumably the being referred to in Genesis 1:24-31, first appeared in Europe about 50,000 years ago; but modern man may have been developed in our present form more than 100,000 years ago. Nevertheless, modern man was the last of the animal species to develop into their present form. Evolution that occurs now seems to involve only the mixtures of various animals that can interbreed, with their descendants competing for survival. Only the "fittest" for the environment they are born into survive.

Creation and Development of Humans

Plural God:

Genesis 1:26 "And God said, Let us make man in our image..."

The implication of the pronouns "us" and "our" is that God is plural, that there was more than one being involved in the creation of human beings. Christians believe that the being called Jesus Christ was present and was included in the term "us." Either all or almost all religions, including monotheists, that teach some form of theism teach that there are very many spiritual beings that provide some kind of service for the Supreme Being or for mankind. Their names are angels, saints, lesser gods, and/or Son(s) of God. Trinitarians believe Christ *and* the Holy Ghost were together with the Father during the creation period; angels probably existed before creation of mankind and were there, too.

The major implication from the use of plural pronouns is that the term *God* refers to a group of beings. Therefore, when the Bible uses the term *Father God,* it probably means only one of the beings. (Although, "Father God" might refer to the corporate God.) It seems that the Bible teaches a modified form of polytheism. Christians claim that *The Son of God, Jesus Christ* is one with *God* and is actually part of *The God.* This relationship is not incompatible with the concept of polytheism. I do believe that the Son of God lived in a human body that physically died 2100 years ago. If this son is part of God, then wherever the Bible uses the word *God* it is referring to the corporate spiritual being; and the Father, Son, and Holy Ghost are members of the group; but they can act as individuals. Probably, there are more Sons of God that have individual responsibilities. (This subject is discussed in more detail later in this book.)

GOD'S FORCE

The Star Wars series used the term "force" to indicate some unseen strength that was available only to select people. That is not quite true to life. The force that made all the objects in the Universe is present in each one of us—in each living being. It is a natural consequence of being alive.

Force in technical terms refers to the amount of effort used to move, or stop, something. (The basic equation is F = M x A: FORCE ==MASS times ACCELERATION. This includes *de*celeration.) It takes effort—force to accelerate or change the position or movement of an object, as stated in the laws of physics that bodies remain at rest or in motion until some outside force acts on them. Energy is the term used to define the amount of work that can be accomplished with available resources such as gasoline or stored electricity in batteries. For example, an automobile engine might be capable of delivering a couple hundred horsepower; but the length of time that horsepower can be delivered is determined by the amount of energy available: in most cases, this means the chemical energy stored in a fuel tank. Creating the Universe took an incredible amount of energy.

Physicists and astronomers have sought vainly to discover a single force that unifies the four fundamental forces. This means *the single force that created the four fundamental forces, the physical laws, all things, all beings, and has continued to control the progress of evolution.*

Theists believe this force is the Spirit of God. So do I. The rest of this book follows this concept that there is a Supreme Being somewhere that is responsible for creation of the universe and for the evolution of mankind. I believe this Being continues to be concerned about the spiritual welfare of human beings.

The single force that made the Universe is present in every living being—in a very small way, of course. It is this force which distinguishes being alive from being dead. The Bible tells us we are made in the image of God; therefore, we must have a small amount of this ability, too. This is the force in paranormal experiences, such as Extra-Sensory Perception (ESP). Humans and evidently some animals, seem to have ESP abilities, which represent in an infinitely lesser degree what the Supreme Being with supreme powers used to create the Universe and continue the development of human souls. The Bible mentions the *"breath of life,"* implying that this was the force of God.

> **Genesis 2:7** And the LORD God formed man of the dust of the ground, and breathed into his nostrils the **breath of life**; and man became a living soul.

Evidently, before the Creator breathed life into man, Man was not or did not have a soul, at least not a "living soul," It is not clear whether there is a difference between "living" soul and just plain "soul." This could mean that only humans have a living soul, which was installed only in the bodies of animals that evolved to the level compatible with having a *living* soul.

The above verse was taken from the King James Version of the Bible. Some other versions say "*living person*" or "*living being,*" instead of "*living soul.*" If the word *person* is correct, it could mean the same as soul, because one definition of the word *soul* is the same as for *person.* The word *being* includes all living animate things, and makes Man merely one of the other animate things. If either one of those is the correct interpretation and excludes the idea that Man is special, my suggestion may not be correct. However, the preceding chapter of Genesis (Gen 1:27) states that Man should "*have dominion*" over all other living beings. Therefore, humans are a special type of being with special responsibilities. I believe God gave us some kind of extra or different characteristics and responsibilities, because we are a different kind of being.

Man's Responsibility

God's force has driven and controlled physical evolution as new types of beings were created; and this force was responsible for extinction of types of beings that were no longer suitable for the divine purpose. Spiritual evolution now is more important to the future of mankind. In order to evolve spiritually, we must learn how to use the force of God within us. Primitive religions, including witchcraft, were necessary and "good" during their times. However, today there is need for a more elevated, more evolved concept of God and of Man and our relationships with God and other humans.

The spiritual information in the Bible is adequate for modern man; although much of its message has been distorted and perverted. Even though the Bible contains spiritual guidance from about 4000 years ago, few people have learned the lessons recorded there. Individuals or societies that have the knowledge and refuse to implement it can expect that God will correct them.

The Old Testament records the history of the Jews and how they reached a relatively high level of spiritual development that was higher than any competing society. Periodically, they backslid into lower levels of spirituality and refused to advance. Then God corrected and molded them to higher levels through persuasion, prophecy, and persecution. Just before the Hebrew nation disappeared, Jesus of Nazareth attempted to revive them once again without much success. However, he did indoctrinate a few thousand disciples that brought the message of truth to the pagans surrounding the Jews.

The majority of Christian groups are not much different from the pagan religions that Christianity supplanted. Yet, Christianity is an extremely advanced religion. Some religious groups use the term "progressive revelations," which Christians believe, without realizing it, is the way their denominations developed. The history of Christianity has demonstrated progressive revelation within itself as new orders and sects were developed. Progressive revelations does not mean that the message of God should be modified to fit modern desires, selfishness, greed, and lusts. It also does not mean that new groups are "better" than previous or other religions; even though, in general, Christianity probably offers more opportunity for spiritual growth than most others do. The reason we have had progressive revelations is because *we* have progressed spiritually.

EVOLUTION

EVOLUTION OF MANKIND

If evolution was the process by which modern animals have developed, it was a very complicated and lengthy process. Assuming that is true, there were many steps involved in reaching the present level of beings on this planet. Each step in evolution required a permanent change in appearance and characteristics from preceding conditions; and all descendants of those animals had to inherit those changes.

If there is such a process as evolution, how does it occur? The geological and paleontological evidence (fossils) indicates animals that might have been ancestors of modern animals changed considerably during the previous millions of years. It appears that some of those fossils might belong to ancestors of human beings. We are not certain which fossils belong to beings that were our ancestors or merely relatives of our ancestors. All that we know for sure is that later fossils show more traits similar to present day humans than earlier ones. Perhaps none of the fossils represents ancestors of modern Humans. Nevertheless, the evidence is pretty strong that Humans evolved from some kind of ancient animal.

According to the paleontological evidence, the present general appearance of mankind was fully developed more than 50,000 years ago, and possibly more than 100,000 years ago. We have little evidence of their mental or spiritual stature, except they dabbled in art, quite beautifully. They seem to have shown respect for the dead, demonstrated by burial practices, indicating, perhaps, belief in an afterlife.

Even if one believes that God created humans instantaneously in their present forms, one can still believe that the Creator used animals that had evolved during millions of years into suitable life forms to "make" human beings only 6-8000 years ago.

The Principles of Evolution

The principles of evolution are: (assuming the most basic form of life has already developed.)

Living creatures must be able to reproduce and multiply.

All creatures that reproduce have descendants that reproduce and resemble their ancestors.

From time-to-time, something happens that interferes with accurate reproduction. This is called "mutation," which could be caused by damage to or alteration in reproductive characteristics.

Some of the creatures resulting from these inaccurate reproductions survive.

Some of these altered beings or things reproduce their individuality continuously.

If the altered beings remain in the same vicinity and compete for the same resources, they must be more capable of surviving and reproducing than other beings that compete for the same resources. Relatives that are less suitable for the environment *must* die and become extinct or else there will be no room or food for the "mostfit" specimens. (Evolutionists call this "natural selection," a useful term, but it applies to only part of the evolution process.)

If, for some reason, the new forms do not compete for the same resources, they could remain in the same locality as their relatives. In some cases, the new forms move away from their relatives and discover an environment that is not suitable for descendants of their common ancestors and flourish there.

The appearance of sexual differentiation speeds up the process of evolution by introducing the powerful recessive genes.

These new surviving creatures are more complex or more "improved" than their ancestors and related creatures.

The above process continues, providing new forms indefinitely.

Living forms that can obtain nourishment from a wide variety of sources are more likely to survive and flourish than those that can use only a single or limited type of sustenance. In addition, those that can survive in a wide variety of climates and environments are able to expand into areas unfit for other, less adaptable forms. The development of memory and reasoning powers is a great advantage and most animals existing today seem to possess these characteristics in some degree. Humans have demonstrated superior ability to flourish with many types of food and survive in many climates, and evidently have the best memory systems and reasoning powers of any other life form on Earth. (Language and writing supplement our memories. Being able to transmit acquired knowledge to succeeding generations is a great advantage—when successive generations accept the knowledge.) Consequently, humans are, at this time, the dominant beings on Earth. However, our superior position and our continued existence demand that we care for our environment and respect other life forms that do not threaten our survival. We must also accept our responsibility to continue evolving.

Male and Female Parents

In order for these steps to proceed rapidly (geologically speaking), it is necessary to involve two beings in the process of reproduction. Without participation by two separate beings, normally one being male and the other a female, evolution would proceed only with multiple, favorable mutations, a rare and undependable form of modifying descendants. Somewhere in the evolution process, after the basic concept of reproduction was developed, the concept of two separate sexes was created. (Again, as in generation of the basic life form, a special touch by God was required.) Only the simplest forms of life can reproduce without participation of another being. Some more complex life forms, such as snails, are able to perform functions of male and female beings, but still, two separate beings are required. None of the higher animals has this characteristic. With two sexes participating, the process of evolution can develop new species much more rapidly than self-fertilized reproductions.

The ability to reproduce is not an automatic characteristic of complex organisms. For example, viruses cannot reproduce without the presence of another organism that can reproduce. Viruses have a unique attribute. They can join with another, independent cell and these joined cells reproduce to become different forms than either one was before being joined—usually resulting in cancerous growth. Some plants, most likely as a result of tampering by mankind, do not have fertile seeds and can be multiplied only with cuttings of twigs or roots.

Favorable mutations in the male reproductive system are more likely to occur than in the female. Sperm is smaller and more fragile than the female egg. Since the quantity of sperms is extremely large, and only the most vigorous sperm succeeds in reaching the female ova, it is more likely that mutations in the sperm will result in successful births of viable offspring. Damage to the fertilized female egg, or failure of the egg to grow properly, usually results in a miscarriage. In most animals, males compete for females; and the most powerful (or richest) males service multiple females. This means that it is likely that there will be more offspring altered from their ancestors if the mutation occurs in a male.

Not only was the development of two sexes advantageous for evolutionary changes, but it seems that the appearance and functions of each of the two sexes had to be unique. Females of most species seem to be receptive to males only at the proper time for reproduction to take place. Since females normally are not always receptive to males, the males had to be ready at any time to take advantage of the female's sudden receptivity.

Females are offspring producers and caretakers. The duty of males is primarily fertilization of the female. Female spiders kill and eat the males after fertilization; and spiders continue to multiply without other influence from male presences. The female spider abandons the eggs after they are placed in a hospitable location. Females of larger animals are not cannibals; and they take care of their young. Female mammals provide milk for their offspring, usually until the young are capable of caring for themselves. Since females are more vulnerable to attack during the period of carrying and caring for offspring, males had to develop the instinct and ability to protect them. This led to development of the desire of females for the largest and strongest males: which led to competition between males for the honor of servicing a female. (There are exceptions to this statement: female hyenas, for example, are bigger, stronger, and more aggressive than male hyenas are.) In humans, this competitive urge can and has caused needless and senseless strife between individuals and nations.

If we accept the theory of evolution currently taught in public schools (without God), what was the process that started things to live? Life is defined as the ability to take nourishment, grow, and reproduce. What caused the first spark of life to burst into being? Lightning striking a pool of ammonia? Where did the ammonia come from? How about the lightning? Crystals can grow under certain conditions, but they do not reproduce in the way that develops into fertile offspring. What caused these early life forms to start reproducing? The concept of "accidental" or spontaneous development of life is extremely difficult to support and defend.

Smart Apes

The following scenario is my interpretation of current theories about the evolution of modern human beings. I used only a few steps in the process of evolution to show how the process actually works. In this case, I assume the existence of a band of monkey-like primates living in a forest somewhere, probably in Africa, and breeding in the manner of most primates. These animals had already evolved from the earliest life forms. We know monkey or ape-like primates existed millions of years ago—at least there is fossilized evidence they existed. I assume, whether actually the case or not, that evolution proceeded from the first generation of life up to these creatures and on to the smart apes in pretty much the way I will describe.

Sexual activity was promiscuous and indiscriminate, dominated by one male. Usually the dominant male had the greatest amount of activity but any male might participate in breeding. At some time after he was conceived, something happened to at least one male that mutated the production of his sperm. The change resulted in a recessive gene that was distributed to females and passed on to subsequent generations. The recessive gene lurked in the offspring for several generations without demonstrating itself. Later, a male with the recessive gene mated with one or more females that also had recessive genes; until finally, several young ones were born that were the result of the initial mutation. The offspring of this(these) mating(s) did not have the dominant genes that were replaced by the recessive gene.

The peculiar effect of this mutation was to delay the maturation of those affected by the gene. This generation was more like human babies than offspring of other members of the band. They never developed the same amount of hair as their parents. They took longer to grow up and their brains continued to develop for a longer time, resulting in more intelligence. Their legs were longer and their bodies were heavier. Because of this, many places in the trees were not sturdy enough to support them so they spent more time on the ground than the rest of the band. (These changes might have occurred suddenly in one generation or in many generations.) Consequently, they had to find food sources other than what was available to their relatives in the trees. They started killing and eating smal animals on the ground.

Because of their feeding habits, their peculiar appearance, and their strange odor, mating with the original primates was not tolerated by either group. The new breed of apes separated into a new band that wandered out of the forest. When they went into the plains, they were met by predators larger than they were that looked on the new beings as new sources of food. This development would have been repeated many times until a band of mutated monkeys was able to survive and multiply in the strange environment and conditions.

During the first few encounters with the predators, they observed that the vicious animals used their front legs in a manner that suggested to the smart apes they could make an artificial arm out of a tree limb and use it as a club. The next time they met one of these predators they used clubs and defeated the animal, feasting on the predator instead of the other way around. These pre-humans, with their superior intelligence and their weapons, were soon masters of the plains, as long as they were in sufficiently large groups. At first, there were few of these creatures, but they were able to survive and reproduce for many generations, perhaps never having more than a few dozen or so in a group. It may have been thousands of years before there was more than one group. Probably other mutants were born, and also spread away from the first group. Eventually, there were many of these separate groups, each continuing to evolve in its own way, because of food, environment, and competition from other animals. Eventually, a few members of these varied groups mated with each other and developed new forms of beings that gradually changed by continuation of the same process that has ended in development of human beings.

There may not be "missing links" between altered forms; because changes might have been too drastic to identify relationships with previous ancestors. On the other hand, the missing links, if there were any, might have involved only a relatively few beings until they continued evolving and multiplying when there were enough of them to leave enough physical evidence to develop into fossils that are thought to be ancestors of human beings.

During the course of our evolutionary climb, humans acquired and still retain aspects of our lower animal nature, especially in relations with other humans. The desire to dominate territories and fear of strangers, especially those who seem to be different from our families keep surfacing in our relations with other humans. This demonstrates itself in what are called prejudices, in bigotry, and in severe cases, annihilation of outsiders, especially competitors. (The purpose of religions should be to help us overcome this animal nature. Unfortunately, some religions do not do this, and instead, teach members how to behave with animal natures and how to overcome others.)

Humans and Dogs

During the course of Man's evolution, he has acquired some friends and many parasites from the other beings that shared the Earth with him. The dog is the most valuable friend and most in harmony with the ways of modern mankind. Some anthropologists think men (or women) merely happened to find motherless wolves and adopted them, domesticated them, raised them as pets. Others think that the ancient wolves started lurking around camps of humans, feeding on garbage that accumulated in waste piles and the vermin that populated the scrap piles. The reduction in offensive odors and unpleasant vermin was of value to the humans and the man-dog relationship commenced.

Either one or both of these explanations might be true. However, I think that the first relationship of canines with beings that were ancestors of modern man was probably a matter of the smart apes living off remains of animals left by canines, probably wolves or wolf-like canines, finishing the leftovers of a kill. Later generations of smart apes joined into hunting efforts with the wolves, something like the way dogs are still used in tracking and finding animals and fowl. Men saw the way that wolves were able to find prey with their superior senses of smell and hearing and then herd the prey into compact groups, resulting in the killing of game. At first, humans followed the wolves and perhaps at first men merely consumed what the wolves left and later they learned to chase the wolves away from prey killed by the wolves. Men may have learned how to hunt from watching wolves. The first spears may have been the teeth of wolves attached to sticks, or were merely imitations of the teeth. Wolves were more efficient hunters than men; but men were better able to plan for the future and store food for leaner days. Therefore, in the lean times, wolves started lurking around camps as mentioned earlier. Eventually something like the modern dog was developed and Man and canine developed a mutually profitable partnership.

Associating with wolves may have contributed to the present physiological condition and the social structure of mankind. Chasing game and keeping up with the canines supported survival of men with long, nimble legs. Depending on the superior senses of canines discouraged development of senses available in the canines and encouraged improvement of other traits that allowed further development of mankind's unique mental prowess.

The social structure of mankind parallels that of wolves more closely than it does that of non-human primates. At first, humans were probably more herbivorous than carnivorous; but when mankind began to hunt with wolves, they also started to depend more on flesh than on plants. Animal flesh is a more efficient source of nourishment than vegetation, which gave mankind more time to use his head instead of his legs.

Instead of wandering around gathering plants, humans learned to stay in one place for long periods. When it was necessary to leave for fishing or hunting, females remained to care for the young the way wolves do. By staying in one place, humans (probably females) who stayed at home learned how to propagate and grow plants providing a dependable source of food and increasing the interdependency of man and canine. The stockpiling of food is a common trait of canines, a trait that may have inspired mankind to do likewise more efficiently. Burial of dead humans may have also been copied from the digging actions of canines.

Conclusions

The conclusion I draw from the preceding discussions is that a supernatural force was necessary for creation and development of the Universe, the Solar System, Earth, and mankind. I believe this force originated in what Christians call "God."

Based on this belief, and what I have read in the Bible, I assume God has the ability to create some kind of energy probably something like what we call electricity. Humans, made in the image of God, can generate electrical energy while thinking or using our brains. If we can do that on a very small scale, I believe God can and did do this on an infinitely greater scale. As discussed earlier, the famous equation, (e=M times c squared) means that material can be created directly from energy, the reverse of energy being generated by nuclear reactions. Therefore, I believe this equation proves the possibility of a celestial being, infinitely greater than human beings, with infinite mental energy, causing the creation of the Universe—and the living beings that inhabit our portion of the Universe merely by using something like brain power.

HISTORY OF MANKIND

Evolution describes the way God created humans. Whether this was accomplished in a day, over a few thousand years, or in a few billion years, is irrelevant. In order for the Earth to be suitable for human beings, it was necessary for the process of creation to go through or appear to go through the steps that are displayed in the fossils on our planet and in astronomical observations. If our physical bodies were not similar to the lesser animals, we could not use any of them for food. Therefore, I think God wants us to understand the processes of creation and evolution.

Evolution of humans that made us different from the forms that resemble humans gave freedom from environment. Evolution in animals without ability to learn from ancestors causes dependence on the environment. Dumb animals (dumb meaning unable to communicate complex messages) can transmit only inherited physical characteristics and a few acquired social relationships to subsequent generations. Modern humans have acquired knowledge from the recorded efforts of many earlier people from many different cultures. (Recently it was noticed that Chimpanzees that learned how to communicate with humans passed these lessons on to their offspring—perhaps leading eventually to an "educated" group of Chimps.)

The most important thing to remember is that humans have evolved to a condition making us different from other animals that have less intellectual capacity. We must continually remind ourselves that we could revert to the animal nature if we do not continually and consciously think and behave like humans.

The latest step in the evolution of modern man was the development of religion, which marked the beginning of man's freedom from his animal nature. Almost all religions teach that humans can draw close to God; and we can live forever and ever in some manner. The manner and conditions of that eternal existence depend on how we live our lives on Earth.

The lessons of history are included in documents and oral

traditions. It is the responsibility of every individual to have some concept of what happened in the past; and the more knowledge of the past, the better it will be for the individual and for the entire society the individual is part of. It cannot be said too often: we must learn from history or we will make the same mistakes that led to the downfall of societies that at one time dominated large areas and great populations. We, at this time, can continue to evolve socially and spiritually, we can stagnate and die, or we can revert to the animal characteristics that still control so many people and societies, including some very large religious groups.

History

Strictly speaking, the term "historic" applies only to the documented period of human existence on Earth. Historians differ on the period of time this really covers. Usually this means a continuous, sequential record of events and people from about 4000BC. Everything that happened before that was prehistoric. However, there are records of humans during the so-called prehistoric time. Legends and oral anecdotes passed from ancient times provide quite a bit of information about earlier times. Even these records, however unreliable we think they are, only go back no more than about fifty thousand years with decreasing reliability the farther back the evidence, pictures, and stories take us. Before that, we have stone and metal tools and weapons, shards of pottery, ornaments, and pictures on rocks, which give us a hazy understanding of the earlier humans.

As the Earth developed, at first molten material, called igneous material, and water in the form of steam erupted from within the mass of space debris that was accumulating into the planet Earth. The igneous material solidified and separated from the water. Water and dry land traded places many times. Dry land became eroded by rain and wind, resulting in debris being carried to the water. When lakes and seas were in place, this debris and remains of animal life were deposited on the bottoms. Over many centuries, this material became compacted, turning into rock, including fossilized remains of animals and vegetation. Then when the water receded, or the bottom of the pool rose above the water level, this material existed as a layer of rock. While this layer of rock was above water and exposed to the atmosphere, it eroded, repeating the same process, with the accumulation of identifiable layers, like a layer cake, called sedimentary rocks. The rocks eroded and decayed along edges, leaving mounds of solid material as mountains or hills. Occasionally molten material intruded into the sedimentary rocks. Some of the rock contained radioactive material. Geologists can determine the age of sedimentary rocks by the decay of radioactive materials such as radium and uranium when these materials are in or associated with the rocks. The age of sedimentary rocks is also determined by comparison and depth of material and fossils within the unknown rock against those found in known rocks.

Once the age of a level of sedimentary rock is calculated, ages of other layers of rock above and below that level can be estimated. This is not foolproof because some lower, older layers have been folded over and appear to be newer than the layers below them. Archeologists and geologists work together to determine the age of the Earth and various kinds of fossils. When the age of a certain kind of fossil has been determined, it is assumed everywhere that kind of fossil appears, the rock containing the fossil is the same age as other rocks where the age of that fossil is known. It is not necessarily true that similar fossils at a different location originated about the same time: evolution of animals and vegetation might not have progressed at the same rate in other locations. However, this process is sufficiently accurate to help us understand what has happened in the past.

The oldest rocks containing radioactive material have been determined to be about two billion years old. (Hindus believe the Earth is about that age.) Because the measured amount of decay could only begin after material solidified, Geologists believe the Earth is older than that by at least two times that length of time.

The complete history of Man starts before there was any life of any kind on Earth, with the beginning of the Universe and the Solar System. The entire record of creation and growth of the Universe, the Solar System, and mankind, as it appears to us in what we believe is physical evidence, is condensed in the following charts. I appropriated material from recognized historians, other experts, and from sensationalists that I believe have some credibility. Dates are approximate and they vary quite a bit from expert to expert. The charts represent my best guess about which dates are most likely the correct ones and in many cases the dates shown bracket the guesses of experts.

CHART NO. 1

NATURAL HISTORY OF THE PLANET EARTH
Formations and Related Life Forms

From as much as 20 BILLION YEARS or as little as 4.5 BILLION YEARS AGO:
The Universe begins with an explosion, called the "big bang," converting energy into material. Sub-atomic particles are flung with tremendous speed into the empty area surrounding it. This area is probably absolutely nothing, even less than the area called "space." Some astronomers think space was created at the time of the big bang.

Perhaps as much as 6.5 BILLION YEARS or as little as 4.5 BILLION YEARS AGO:
The Solar System begins to form. Objects from other parts of space in the form of gases, dust, comets, meteors and small planets collide with one another, commencing formation of larger particles that eventually become the Sun and the planets, including Earth. At first, Earth is gaseous and liquid, not solidifying until its weight causes it to compact and the surface cools. Some of the original material fails to be attracted by or to join a planet or satellite at first and continues to maintain independent orbits. This includes comets, asteroids, and meteoroids with orbits that allow them to occasionally come close enough to planets, satellites, and other objects to bring them into collisions with them.

Since observable and measurable objects are not available on Earth from this period, the most significant clue to the age of this period is from analysis of meteorites and other debris assumed to be from the original formation period that continue to strike the Earth. However, one problem in determining the age of these meteors is that some of them probably at some time during their existence traveled at or near the speed of light, which means they may actually be older by Earth time than they appear.

> No traces of life from this period can be found anywhere, although some meteorites have indications that simple molds might have been developed.

4.5 BILLION YEARS to 3 BILLION YEARS AGO:
Earth's crust begins to form. Space debris continues to add solid material to the mass that is forming into the Earth. Surrounding this mass is a cloud of dust and gases at first. Near the end of this period, Earth's surface is mostly solid material. Liquids begin to precipitate from the

gases. Other liquids and gases separate from the molten material.

No traces of life found in these rocks.

3 BILLION YEARS to 2.0 BILLION YEARS AGO:
Igneous rocks continue to form from volcanic action pushing molten and gaseous material to the surface. Water and other liquids fall as rain and flow in channels causing erosion of igneous rocks, commencing development of sedimentary rocks where the liquids settle in basins. Atmosphere at first part of period is composed mostly of carbon dioxide with traces of other gases.

Traces of blue-green algae, the first indication of life on Earth, are deposited. The algae convert carbon dioxide into oxygen, commencing development of the present atmosphere.

2.0 BILLION YEARS to 1.7 BILLION YEARS AGO:
Sedimentary layers of rock containing fossils of life forms are established. Lava flows continue, covering and/or intruding into the sedimentary material.

Microscopic organisms: bacteria and probably viruses develop.

1.7 BILLION YEARS to 1.2 BILLION YEARS AGO:
Great layers of sedimentary rocks form, some with granite intrusions.

Worms leave traces. More algae forms develop.

1.2 BILLION YEARS to 800 MILLION YEARS AGO:
Thick deposits of sedimentary rocks form. Traces of glacier formations are found.

Multi-cellular life forms develop.

800 MILLION YEARS to 700 MILLION YEARS AGO:
Intense volcanism occurs. Great deposits of iron are deposited in Lake Superior area. First coal layers deposited.

Life form diversification explodes.

700 MILLION YEARS to 550 MILLION YEARS AGO:
Climate is mild. Tropical climate exists in North America. Products of erosion create layers of sedimentary rocks mixed with fossils.

All animal and plant life that exists lives in seas. Animals with shells and skeletons develop.

550 MILLION YEARS to 435 MILLION YEARS AGO:
A large land mass containing what are now the separate main land masses begins to break up, commencing continental drift. Igneous rock (lava) formation slows.

Mollusks, corals, and fishlike vertebrates develop.

435 MILLION YEARS to 395 MILLION YEARS AGO:
Widespread limestone, red rock—indicating the presence of atmospheric oxygen, and salt layers are deposited as water retreats from land areas.

Earliest land plants develop. Coral reefs are built up. Sharks appear.

395 MILLION YEARS to 345 MILLION YEARS AGO:
Europe and North America collide, causing mountains to rise. South Pole is in central Africa. (Or, considering continental drift, central Africa may be at the South Pole, if the poles have not shifted.)

Many fish forms exist and dominate the planet. Amphibians are first land animals. Plant life covers planet. Primitive seed plants appear.

345 MILLION YEARS to 320 MILLION YEARS AGO:
Sea covers Mississippi valley. Climate is warm. Vast swamplands with coal-forming sediments are deposited. Severe continental collisions continue to create mountains in Europe.

Fern like trees and primitive conifers appear.

320 MILLION YEARS to 280 MILLION YEARS AGO:
Mountains form in Oklahoma area, Europe, and Asia. Extensive coal deposits formed in central North America.

Insects flourish. First reptiles appear toward end of this period.

280 MILLION YEARS to 230 MILLION YEARS AGO:
Appalachian Mountains and Urals are raised. Glaciers form in southern areas of planet.

Many life forms become extinct. Other forms flourish and evolve rapidly. The number and variety of reptiles increases and amphibians decline. Modern corals appear.

230 MILLION YEARS to 195 MILLION YEARS AGO:
Extensive creation of red beds and other non-marine deposits are laid down. Salt and gypsum deposits build up.

Conifers and cycads are abundant. Dinosaurs and marsupials appear. Flying and marine reptiles are still widespread.

195 MILLION YEARS to 140 MILLION YEARS AGO:
Land masses separate. Drift of continental plates causes folding of crust in North America. Major erosion of mountains begins.

Reptiles dominate land, sea, and air. Early bird appears. First mammals appear. Palm-like plants, ginkgoes, and horsetails appear.

140 MILLION YEARS to 65 MILLION YEARS AGO:
South America and Africa separate. North Atlantic is widening.

Dinosaurs and other large reptiles reach a peak and then disappear. Snakes and lizards appear. Modern trees show up. Flowering plants develop.

65 MILLION YEARS to 50 MILLION YEARS AGO:
Volcanism forms Rocky Mountains and other ranges.

Earliest primates, similar to modern shrew, appear. Most of the modern common plants, including hardwood trees and redwoods spread. Modern birds, early horses, pigs, whales, and rodents appear.

50 MILLION YEARS to 37 MILLION YEARS AGO:
Erosion of mountains begins to fill basins.

Number and types of primates increase. Prosimians appear. North American climate is warm and humid. Placental mammals begin to increase.

37 MILLION YEARS to 23 MILLION YEARS AGO:
Africa and Europe collide, raising Alpine Mountains. Arabia and Africa split at the Red Sea Rift. Seasonal changes start affecting geologic and life form events.

Modern horses, pigs, carnivores, and elephants begin to show up. Cats and dogs begin to evolve.

23 MILLION YEARS to 5 MILLION YEARS AGO:
Volcanism raises mountains throughout world. Climate is colder.

Many new mammals appear and many earlier forms become extinct. Early monkeys and earliest hominoids appear. Extent and size of forests decrease. Grassy plains increase.

5 MILLION YEARS to 1.8 MILLION YEARS AGO:
Volcanism diminishes but it still creates new lands. Climate is cool and dry.

Gorillas and hominids begin to develop about 5 million years ago. Earliest true (bipedal) hominid, Australopithecus, appears about 4 million years ago. Animals and plants begin to resemble modern forms. Rocks that appear shaped by ancestors of humans (tools and trinkets) are from about 2.5 million years ago. Homo Habilis appears about 2 million years ago.

1.8 MILLION TO 10,000 YEARS AGO:
This is the age of glaciers that drift and retreat, causing land changes and creating basins for lakes.

Many mammals (mastodons, mammoths and saber tooth tigers, for example) become extinct. Homo Erectus appears in Europe and Asia at beginning of this epoch, about 1.6 million years ago. Neanderthal man appears about 350,000 years ago. The first Cro-Magnon remains are in Africa and Asia from about 100,000 years ago. Cro-Magnon man appeared in Europe about 35 to 40,000 years ago.

10,000 YEARS AGO TO PRESENT
Climate is warmer. Glaciers retreat. Some great deserts form. Climate continues to vary in irregular cycles between cold and warm, causing advance and retreat of glaciers. Glaciers gouge deep, semicircular valleys as they advance and deposit debris as they melt and seem to retreat.

Modern humans flourish. Modern civilizations begin. Extinction of many early forms and species continues.

CONTINUED DEVELOPMENT OF MANKIND

Ancient Man

Rocks in the shape of tools and weapons, apparently fabricated by beings that probably were ancestors of modern human beings, have been found in sedimentary rocks that were laid down about 1,500,000 years ago. These items were simple and crudely made. Items found in more recent layers of rock demonstrate more sophistication, specialization, and craftsmanship, as the apparent time of their creation gets closer and closer to our time.

Cultures of prehistoric humans are named for the predominant material used as tools and weapons. The oldest culture is named PALEOLITHIC PERIOD, old stone age, although bone was also used. There were two more stone ages: MESOLITHIC PERIOD, middle stone age, and NEOLITHIC PERIOD, new stone age when polished stone and ceramics were used. The CHALCOLITHIC, OR COPPER AGE, started with the discovery that metal was available and made good tools. (Copper tools were used for much of the ancient Egyptian sculpture and architecture.) The BRONZE AGE was next, starting when humans discovered that mixtures of copper and tin resulted in stronger materials. The IRON AGE is the most recent age named after materials used for tools and weapons, commencing when humans learned how to melt and form iron. Dates for the beginning of the different periods and ages vary from place-to-place. The cultural changes appeared at different times in various places, indicating ideas and materials for the various fabrication activities probably migrated with people as they moved from the places the ideas developed and the materials were first used.

Modern Man

The current period has had several other names, such as the INDUSTRIAL, ELECTRONIC, and COMPUTER AGE, depending on how recent a time is being considered. Civilization did not progress smoothly from one age to another: some societies self-destructed—because of environmental abuse, over-population; or internal strife; others were overcome and destroyed by invaders; some migrated and individual identities were lost, and others were lost because of geological or climatic changes that obscured the remains of their cultures.

Chart 2 lists the major cultures and changes that probably occurred. Some of the comments are considered to be legends. However, the legends might be based on actual events.

CHART NO. 2

Development of Modern Civilization

B. C. Dates	Cultural or Geologic Activity
C50,000	Cro-Magnon humans appear in Europe.
C50,000	(legend) Civilization of Mu starts.
C40,000	First humans might have appeared in N. and S. America.
C15,000	(legend) Mu has population of 64 million.
C12,500	Humans have definitely appeared in the Americas by this time.
C11,400	(legend) Mu destroyed.
C11,500	Glaciers start to retreat in North America and Europe.
C11,500	Positive evidence of humans in the Americas.
C10,500	(Plato) War between Atlantis and Southern Europe.
C10,000	(legend) Atlantis, population 30 million, destroyed.
C10,000	Ancestors of Iberians, Egyptians, Celts, Greeks, and Semites appear in Europe and Mediterranean, replacing or blending with Cro-Magnon.
C10,000	Herds of mammoths are inundated by mud and quickly frozen in permafrost in Siberia. Many large mammals become extinct.
C2800-1900	Pyramids built in Egypt.
C900	Etruscans appear in Italy.

LIFE, DEATH, AND RESERRECTION

It appears that, according to the Bible, and modern definitions, and as described earlier, the soul is the part of a human being that contains an individual's personality. Most theistic religions teach that the soul can be eternal, and many believe it can die.

The Lord God told Adam that he would die in the day that he ate of the tree. Quite obviously, Adam did not physically die—he and Eve lived many years and evidently had many children. The serpent said they would not surely die. Therefore, it seems that the serpent was more truthful than the Lord was—unless they were talking about two different kinds of death. That might have been the situation. The souls of Adam and Eve could have died, or would die when the bodies died, even though their bodies survived for many years after being expelled from Eden. On the other hand, God might have meant they would eventually experience physical death. (We do not know how long the day was that God was talking about.) Satan would have meant they would not die immediately. The first three passages below from the Bible tell of the warning of God and the events that followed the transgression. In the verse from James, he states that a soul can be saved from death, implying that the soul will die if it is not saved.

> **Genesis 2:15-17** And the LORD God took the man, and put him into the garden of Eden to dress it and to keep it. And the LORD God commanded the man, saying, Of every tree of the garden thou mayest freely eat: But of the tree of the knowledge of good and evil, thou shalt not eat of it: for **in the day that thou eatest thereof thou shalt surely die**.

> **James 5:19,20** Brethren, if any of you do err from the truth, and one convert him; Let him know, that he which converteth the sinner from the error of his way shall save a soul from death, and shall hide a multitude of sins.

The following verses from Ezekiel tell us what humans must do and refrain from doing in order for the soul to live. In the last verse, the soul appears to be equated with the personality or ego by using the personal pronoun. The verse from Isaiah states the same message in more positive and comforting terms.

> **Ezekiel 18:4-9** Behold, all souls are mine; as the soul of the father, so also the soul of the son is mine: **the soul that sinneth, it shall die**. But if a man be just, and do that which is lawful and right, And hath not eaten upon the mountains, neither hath lifted up his eyes to the idols of the house of Israel, neither hath defiled his neighbour's wife, neither hath come near to a menstruous woman, And hath not oppressed any, but hath restored to the debtor his pledge, hath spoiled none by violence, hath given his bread to the hungry, and hath covered the naked with a garment; He that hath not given forth upon usury, neither hath taken any increase, that hath withdrawn his hand from iniquity, hath executed true judgment between man and man, Hath walked in my statutes, and hath kept my judgments, to deal truly; he is just, **he shall surely live**, saith the Lord GOD.

> **Isaiah 55:2,3** Wherefore do ye spend money for that which is not bread? and your labour for that which satisfieth not? Hearken diligently unto me, and eat ye that which is good, and let your soul delight itself in fatness. **Incline your ear, and come unto me: hear, and your soul shall live**; and I will make an everlasting covenant with you, even the sure mercies of David.

In Genesis 35:18 below, the soul was "departing" from the body. The meaning of the term *"the soul was in departing"* was explained and emphasized by the parenthetical *"for she died."* Evidently, this means that the soul went somewhere; but where is not explained, except that it left the physical body.

> **Genesis 35:16-19** And they journeyed from Bethel; and there was but a little way to come to Ephrath: and Rachel travailed, and she had hard labour. And it came to pass, when she was in hard labour, that the midwife said unto her, Fear not; thou shalt have this son also. And it came to pass, as **her soul was in departing,** (for she died) that she called his name Benoni: but his father called him Benjamin. And Rachel died, and was buried in the way to Ephrath, which is Bethlehem.

The Final Resurrection

At the time Jesus of Nazareth was ministering to the Jews, many people (the Sadducees) did not believe some part of a human being survives death of the body. Many others did believe. They apparently believed that the physical body would be revived at some time in the future and the soul could resume abiding in that body. However, Jesus said that after the resurrection those that live will not have the same kind of body that we think of as a human body. This statement is in the following passage. He also states that Abraham, Isaac, and Jacob were alive at the time he was speaking "*God is not the God of the dead, but of the living.*" Evidently, these men were in Heaven in a form something like angels, or they had been reborn in other bodies and thus living somewhere on Earth.

> **Matthew 22:23-33** The same day came to him the Sadducees, which say that there is no resurrection, and asked him, Saying, Master, Moses said, If a man die, having no children, his brother shall marry his wife, and raise up seed unto his brother. Now there were with us seven brethren: and the first, when he had married a wife, deceased, and, having no issue, left his wife unto his brother: Likewise the second also, and the third, unto the seventh. And last of all the woman died also. Therefore in the resurrection whose wife shall she be of the seven? for they all had her.
>
> Jesus answered and said unto them, Ye do err, not knowing the scriptures, nor the power of God. **For in the resurrection they neither marry, nor are given in marriage, but are as the angels of God in heaven.** But as touching the resurrection of the dead, have ye not read that which was spoken unto you by God, saying, I am the God of Abraham, and the God of Isaac, and the God of Jacob? **God is not the God of the dead, but of the living.** And when the multitude heard this, they were astonished at his doctrine.

The gospel of Luke repeats this incident, but adds the statement that only those who "*shall be accounted worthy to obtain that world,*" will not die anymore and are the children of God, equal to the angels. The time of the resurrection is not stated here, although in John 11:24 Martha says to Jesus that she expects the resurrection to occur "*at the last day.*"

Luke 20:33-38 Therefore in the resurrection whose wife of them is she? for seven had her to wife.

And Jesus answering said unto them, The children of this world marry, and are given in marriage: But **they which shall be accounted worthy to obtain that world, and the resurrection** from the dead, neither marry, nor are given in marriage: Neither can they die any more: for they are equal unto the angels; and are the children of God, being the children of the resurrection. Now that the dead are raised, even Moses shewed at the bush, when he calleth the Lord the God of Abraham, and the God of Isaac, and the God of Jacob. For he is not a God of the dead, but of the living: for all live unto him.

John 11:23,24 Jesus saith unto her, Thy brother shall rise again. Martha saith unto him, I know that **he shall rise again in the resurrection at the last day.**

In the following passage from Luke, Jesus described the conditions of two individuals in the hereafter. Abraham was somewhere in Heaven. The rich man was in Hell, and he could see Abraham, but *a great gulf* prevented transfer of souls and comfort.

Luke 16:19-26 There was a certain rich man, which was clothed in purple and fine linen, and fared sumptuously every day: And there was a certain beggar named Lazarus, which was laid at his gate, full of sores, And desiring to be fed with the crumbs which fell from the rich man's table: moreover the dogs came and licked his sores.

And it came to pass, that the beggar died, and was carried by the angels into Abraham's bosom: the rich man also died, and was buried; And in hell he lift up his eyes, being in torments, and seeth Abraham afar off, and Lazarus in his bosom. And he cried and said, Father Abraham, have mercy on me, and send Lazarus, that he may dip the tip of his finger in water, and cool my tongue; for I am tormented in this flame.

But Abraham said, Son, remember that thou in thy lifetime receivedst thy good things, and likewise Lazarus evil things: but now he is comforted, and thou art tormented. And beside all this, **between us and you there is a great gulf fixed: so that they which would pass from hence to you cannot; neither can they pass to us, that would come from thence.**

SURVIVAL OF INDIVIDUAL PERSONALITY

Scientific and engineering discoveries have reduced the emotional need for magic by developing methods and tools to satisfy material needs of humans. Canals and reservoirs bring water to parched lands, reducing damage of droughts. Weather, while still not controllable, is better understood with satellites and radar scans of local areas; and changes can be forecast, helping to reduce damage caused by destructive weather. Lightning rods reduce chances of damage caused by electrical storms. Disease is controlled to a great degree by sanitation, medications, surgery, and other treatments based on anatomical studies. After nature came under limited control, and we understood some of the vagaries of nature, an important concern for human beings remained. No matter how healthy a person is or how long the life span, death is an inevitable consequence of being born. Many people, probably most people, believe that some part of an individual survives death of the physical body. Usually this part is called the *soul.*

Conquest of Death

Paul told us that death will be conquered at the end of this era of civilization. Even if it cannot be avoided now, the message is clear that death can be overcome.

> **1Corinthians 15:51-56** Behold, I shew you a mystery; We shall not all sleep, but we shall all be changed, In a moment, in the twinkling of an eye, at the last trump: for the trumpet shall sound, and the dead shall be raised incorruptible, and we shall be changed. For this corruptible must put on incorruption, and this mortal must put on immortality. So when this corruptible shall have put on incorruption, and this mortal shall have put on immortality, then shall be brought to pass the saying that is written, Death is swallowed up in victory. **O death, where is thy sting? O grave, where is thy victory?** The sting of death is sin; and the strength of sin is the law.

Death

Reports in the Bible tell us that Adam and some of the other men living about that time lived for hundreds of years. Descriptions of them make it seem that these were real individuals who actually lived that long. It is possible they did live that long—living conditions in the pre-flood era might have been much better. Mankind might have been more advanced technologically than we believe they were, and they had discovered the secret of long life.

The people that lived such long lives, if they really did live that long, lived before the flood that inundated the world of Noah, which might have been more than 12,000 years ago. (Something happened about that time that caused the ice cap in the North to melt and the Earth began to get warmer and wetter.) After the time of Noah, humans did not live much longer than 120 years, as mentioned in Gen 6:3. The later patriarchs such as Abraham and Moses lived about that long. Somehow, by the time of David about 1000 BC, as shown in Psalms 90:10, humans were doing well to reach the age of "*threescore years and ten* (70)". Until about 100 years ago, the *average* life expectancy of humans was less than 50. Today, the average life expectancy of men and women is between 70 and 80.

> **Genesis 5:5** And all the days that Adam lived were **nine hundred and thirty years:** and he died.
>
> **Genesis 5:27** And all the days of Methuselah were **nine hundred sixty and nine years**: and he died.
>
> **Genesis 6:3** And the LORD said, My spirit shall not always strive with man, for that he also is flesh: yet his days shall be **an hundred and twenty years**.
>
> **Psalms 90:10** The days of our years are **threescore years and ten; and if by reason of strength they be fourscore years**, yet is their strength labour and sorrow; for it is soon cut off, and we fly away.

According to biology experts, most mammals live about 10 times as long as their puberty periods. The puberty period of humans is about the age of twelve. Therefore, humans should live for about 120 years, as mentioned in Genesis 6:3. In some places, until recently, anyway, people did live that long in isolated parts of the world: mostly mountain regions. When modern civilization advanced into these areas, life expectancy in these places dropped closer to the average for the rest of the world. However, the number of people living to be over 100 does seem to be increasing—perhaps merely because there are more people in the world. Every once in awhile, someone lives to be 120 or even a few years more, but this appears to be pretty close to the maximum we can expect.

Humans have agonized over the unavoidable fact of death and the limited duration of life. The Psalms writer expressed this concern about 3000 years ago. The writer seems to gain satisfaction in the equality of death; all who live will die.

> **Psalms 89:47, 48** Remember how short my time is: wherefore hast thou made all men in vain? What man is he that liveth, and shall not see death? shall he deliver his soul from the hand of the grave? Selah.

(*Selah means to pause and think about what was just said.*)

> **Psalms 49:10-14** For he seeth that **wise men die, likewise the fool and the brutish person perish,** and leave their wealth to others. Their inward thought is, that their houses shall continue forever, and their dwelling places to all generations; they call their lands after their own names. Nevertheless, man being in honour abideth not: he is like the beasts that perish. This their way is their folly: yet their posterity approve their sayings. Selah. Like sheep they are laid in the grave; death shall feed on them; and the upright shall have dominion over them in the morning; and their beauty shall consume in the grave from their dwelling.

> **Psalms 37:35-36** I have seen the wicked in great power, and spreading himself like a green bay tree. Yet he passed away, and, lo, he was not: yea, I sought him, but he could not be found.

> **Psalms 39:4, 5** LORD, make me to know mine end, and the measure of my days, what it is; that I may know how frail I am. Behold, thou hast made my days as an handbreadth; and mine age is as nothing before thee: verily every man at his best state is altogether vanity. Selah.

Solomon in his book of Ecclesiastes repeated the feeling that life is *altogether vanity.* Toward the end of his life, he was rather pessimistic about life and death. The following passage from

Ecclesiastes expresses the concern many people have when they reach old age and death seems close.

> **Ecclesiastes 12:5** Also when they shall be afraid of that which is high, and fears shall be in the way, and the almond tree shall flourish, and the grasshopper shall be a burden, and desire shall fail: because man goeth to his long home, and the mourners go about the streets: Or ever the silver cord be loosed, or the golden bowl be broken, or the pitcher be broken at the fountain, or the wheel broken at the cistern. Then shall the dust return to the Earth as it was: and the spirit shall return unto God who gave it.
>
> Vanity of vanities, saith the preacher; all is vanity.

> **Ecclesiastes 1:3, 4** What profit hath a man of all his labour which he taketh under the sun? One generation passeth away, and another generation cometh: but the Earth abideth for ever.

Finally, the statement of David in the following Psalm expresses the feeling of most humans that believe in a Supreme Being.

> **Psalms 8:4-6** What is man, that thou art mindful of him? and the son of man, that thou visitest him? For thou hast made him a little lower than the angels, and hast crowned him with glory and honour. Thou madest him to have dominion over the works of thy hands; thou hast put all things under his feet:

Persistence of Personal Existence

Jesus Christ taught that some significant part and awareness of individuality survives death of the physical body.

> **John 3:16** For God so loved the world, that he gave his only begotten Son, that whosoever believeth in him should not perish, but **have everlasting life**.

The belief in continued existence of individual personalities in a spiritual realm is the most important part of the religion founded on his teaching. Christians believe that the death and resurrection of Jesus proved some part of individuals also will survive physical death because Jesus Christ demonstrated the possibility. Until the nineteenth century, most people accepted this on blind faith. Toward the end of the 1800's, along with the increasing knowledge of the physical world, people began studying the possibility of proving "scientifically" that human personalities survive physical death, and that their "spirits" could communicate with living persons.

People with training in various scientific endeavors began to study random reports by varied individuals of communications with or visitations from spirits. This was the first serious attempt to coordinate what was known about science, religion, and death. The spiritual experiences of humans were studied by scientific methods in a philosophy that is now called Spiritualism or Spiritism. *The Society for Psychic Research* formed and began to accumulate records of these occurrences, which were called *physic, paranormal,* or *Extra Sensory Perception (ESP)* experiences. This subject was discussed earlier in this book.

People, called *mediums,* who seem to receive direct communications from people who have "died," participated in a study, using what was called "cross communication." This involved two or more mediums who were not acquainted with one another and lived in different areas. Mediums that were unaware of reports from other mediums were able to continue or refer to communications delivered previously by other mediums in widely separated places. This was good evidence that the spirit of the dead person was able to continue the message through the new medium. The only link between the mediums was the spirit of the dead person. This method eliminated concern that information supposedly derived from deceased individuals could be the result of telepathy between living humans, clairvoyance, or deceit.

In the first few years of the twentieth century, F. W. H. Myers compiled records of these investigations. They were published in a book, *Human Personality and its Survival of Bodily Death.* Susy Smith revised the original material and the book was reprinted in 1961. The studies reported in the book were conducted as scientifically as any subjective experience can be. The book presents strong evidence that communications between living people and people who had died had actually occurred. In addition, the book discusses related subjects such as ESP, hunches or sudden insights, split personalities, ghostly appearances, and out-of-the-body experiences, as well as communications with those who had "passed."

These studies stimulated interest in ESP and spiritualism. Many people repeated earlier experiments in many places during the twentieth century. Lately, serious study of paranormal abilities seems to have halted. The lack of interest in scientific study of the subject might mean that there does not seem to be any possibility of learning anything new; or in might indicate that humans have become interested only in making their present lives more comfortable. Rumors have circulated that some governments experimented secretly with various forms of paranormal activity, evidently hoping to penetrate the minds of opponents or to develop psychic methods of harming or killing people.

Many people refuse to accept results of paranormal and spiritualistic experiences because of fear, prejudice, or a desire to ignore the entire subject of death. Evidence is strong for believing that messages supposedly from deceased persons can truly be from them; and the characteristics displayed by them "after-death" were the same as those shown while on Earth.

Communications with the Dead

In the following passage from the book of 1Samuel we are told that the spirit of Samuel was recalled from beyond the grave by the woman of Endor to provide guidance for King Saul. The Bible reports this as an actual event. The spirit recognized as Samuel had the same contempt for Saul that the physical Samuel demonstrated before his death. The spirit behaved in the way that Samuel would have behaved; and the message delivered to Saul predicted the future as it happened the next day. Some religious scholars believe the being that confronted Saul was a demon or Satan pretending to be Samuel. Of course, there are scoffers who think this is allegorical and it did not really happen. I do not think a demon would have been able to do that, nor would it have spoken so authoritatively about the Lord. I do not believe that the Bible would use an allegory that uses forbidden activity. The Bible prohibits any attempt to communicate with spirits; but it does not say that it *cannot* be done, just that it *should not* be done. Therefore, I believe the Bible has given us evidence that the soul can have existence separate from the body, it survives death of the body, while retaining its previous earth-bound knowledge and some other characteristics, and it can communicate with humans living in physical bodies.

1Samuel 28:3-19 Now Samuel was dead, and all Israel had lamented him, and buried him in Ramah, even in his own city. And Saul had put away those that had familiar spirits, and the wizards, out of the land.

And the Philistines gathered themselves together, and came and pitched in Shunem: and Saul gathered all Israel together, and they pitched in Gilboa. And when Saul saw the host of the Philistines, he was afraid, and his heart greatly trembled. And when Saul enquired of the LORD, the LORD answered him not, **neither by dreams, nor by Urim*, nor by prophets.** Then said Saul unto his servants, Seek me a woman that hath a familiar spirit, that I may go to her, and enquire of her.

**Urim were gems or stones used by priests to receive messages from God—the stones probably were used like dice to find random numbers or arrangements after being cast on a flat surface. Evidently, the Jews accepted this as revelations from God.*

And his servants said to him, Behold, there is a woman that hath a familiar spirit at Endor. And Saul disguised himself, and put on other raiment, and he went, and two men with him, and they came to the woman by night: and he said, I pray thee, divine unto me by the familiar spirit, and bring me him up, whom I shall name unto thee. And the woman said unto him, Behold, thou knowest what Saul hath done, how he hath cut off those that have familiar spirits, and the wizards, out of the land: wherefore then layest thou a snare for my life, to cause me to die? And Saul sware to her by the LORD, saying, As the LORD liveth, there shall no punishment happen to thee for this thing. Then said the woman, Whom shall I bring up unto thee? And he said, Bring me up Samuel.

And when the woman saw Samuel, she cried with a loud voice: and the woman spake to Saul, saying, Why hast thou deceived me? for thou art Saul. And the king said unto her, Be not afraid: for what sawest thou? And the woman said unto Saul, I saw gods ascending out of the earth. And he said unto her, What form is he of? And she said, An old man cometh up; and he is covered with a mantle.

And Saul perceived that it was Samuel, and he stooped with his face to the ground, and bowed himself. And Samuel said to Saul, Why hast thou disquieted me, to bring me up? And Saul answered, I am sore distressed; for the Philistines make war against me, and God is departed from me, and answereth me no more, neither by prophets, nor by dreams: therefore I have called thee, that thou mayest make known unto me what I shall do. Then said Samuel, Wherefore then dost thou ask of me, seeing the LORD is departed from thee, and is become thine enemy? And the LORD hath done to him, as he spake by me: for the LORD hath rent the kingdom out of thine hand, and given it to thy neighbour, even to David: Because thou obeyedst not the voice of the

> LORD, nor executedst his fierce wrath upon Amalek, therefore hath the LORD done this thing unto thee this day. Moreover the LORD will also deliver Israel with thee into the hand of the Philistines: and to morrow shalt thou and thy sons be with me: the LORD also shall deliver the host of Israel into the hand of the Philistines.

In spite of prohibitions against necromancy, (Deuteronomy 18:10-12) many people practice this art. Some who practice spiritualism are earnestly seeking insight into the meaning of life and death; others merely want the comfort of knowing that they will survive in a recognizable form after death of the body, and some want relief from loss of loved ones. (Necromancy is the practice of conjuring up the spirits of the dead.)

Apparitions

Many people have reported spontaneous and unprovoked experiences apparently involving spirits of persons who had died. A few people have told me of what seemed to be real experiences. The most common is an image of a person presenting itself to a loved one at the moment the person is dying. It is possible that this is not a spirit, but a manifestation of telepathy originating in a person that is aware of impending death. Even if this is merely a form of ESP, it is still evidence of a capability to receive communication through non-material methods not previously or normally available. However, appearances of persons that had been dead for a long time (commonly called ghosts) do not fit the concept of telepathy between living persons. Even so, communications with the spirits of people who had died physically would have to involve telepathy or something like it.

Critics ridicule these reports as hallucinations or hoaxes. Nevertheless, those who have gone through this kind of experience believe something real happened, even if they do not believe in "ghosts." Strictly speaking, the word apparition refers only to visual appearances. However, not all experiences with disembodied spirits involve sightings. Sometimes there is merely a feeling of coldness. Other times it is a strange and unexplainable odor. Sometimes there are strange sounds such as footsteps or voices in an empty hallway. I have had experiences that might have involved disembodied spirits; and several people I know and believe in, have told me of also doing just that.

Charlie's Name: An experience that happened to me involved the sound of glass being struck sharply. Shortly after I was discharged from the U. S. Navy, my dad and I were reminiscing about the old days, discussing our life together and he related events in his life. Suddenly, just as he mentioned the name of one of his close friends who had died many years earlier, a loud cracking sound came from a large, square ashtray on a table near us. We were both startled and stopped speaking. A visual examination showed that the ashtray was still intact, with no visible cracks. "What was that, dad?" I said. He replied, "It happened when I mentioned Charlie's name." Once more, when he uttered the word "Charlie," the ashtray cracking sounded again, less loudly. We were even more bewildered and perhaps slightly frightened. After a few seconds, I said that maybe Charlie was trying to contact us. Once again, the ashtray made the same sound at the mention of the man's name, considerably diminished in intensity this time.

After that, no matter how many times we mentioned Charlie's name, the sound was not repeated. Three days later, the ashtray split in half. The skeptic's comment is that the ashtray was under stress and about to break, just coincidentally occurring when Charlie's name was mentioned. First, glass is under stress immediately after being formed; and because of its peculiar amorphous internal structure, it gradually loses these stresses. Since the ashtray was several years old, there were no residual stresses left from manufacturing. There had been no sudden temperature changes. Second, the cracking sound happened three times only at the time the name was mentioned. So, what caused the noise and the later splitting of the ashtray? My dad and I were convinced that his old friend was saying hello.

His Father's Image: The first person who told me of "seeing" a person who actually was somewhere many miles away was an associate at work. He told me that he was an atheist and did not believe in survival of personalities. Nevertheless, he claimed to have seen the figure of his father at or about the time that the father died. He wanted me to tell him what it meant. Since he had already denied possibility of the answer that seemed most logical to me, I could only tell him I thought his father must have been thinking about him and wanted to assure him that personalities were not restricted to only physical existence.

Her Husband's Face: The following experience was not a vision of a ghost; but it demonstrates the same effect: that of someone desperately wanting to communicate with someone far away. Betty, the woman I loved and married after Geri died, told me about an experience she had during WWII. She was in her home when suddenly she saw the face of her husband, Jimmie. She was frightened and rushed to tell her landlord about the apparition. He encouraged her to record the incident and the time it occurred. Later she found out that at that moment, Jimmie had been standing on the deck of his ship which was participating in the invasion of Southern France when a shell flew by him—so close it made his pants leg move. He was not hurt, but he expected to be as the shell approached him; and he immediately thought about his wife.

The image of someone was transmitted to a loved one far away during an emotionally charged situation. This corresponds with the images of people appearing to loved ones just before or at the moment of death. This sort of thing has convinced many people that it is possible to transmit vivid thoughts to others with only mental efforts. Even though this was not a sighting of a ghost, it is evidence of the possibility that spiritual beings, assuming that there are such things, can also communicate with us.

Reassurance from Grandparents: The father of one of my friends had died. Shortly after the death, his favorite grandchild saw what appeared to be him standing with his wife who had died earlier. They seemed to be solid, live persons. She knew of the deaths but was completely unprepared for the experience; and she was frightened, but quickly realized that the images were meant to reassure her.

Many other people have reported to me that they have seen, spoke with, heard, felt, and/or were touched by "ghosts." I think I also have had several of these experiences. None of my experiences was particularly unusual; although at the times I was certain there was something or someone other than a physical being near me.

Friendly Spirits: Many years ago, a clairvoyant told me that there were four "spirits" of deceased relatives around me. The clairvoyant had no way of knowing the names or descriptions of these relatives, and I had not thought about them for a long time; and three of them had died before I was born. All names and descriptions mentioned were of people that had some special significance for me. One of them was a sister of my mother, someone my mother had said I reminded her of. Another one was the mother of my father. This was a person that my father had said I reminded him of. Both parents had demonstrated paranormal capabilities; and perhaps they sensed the presence of the spirits of these family members. Since I had not thought of any of the people he mentioned for a long time, and I had to go to some old papers to compare the names of two of them, I do not think the clairvoyant could have been reading my mind; therefore, I believe he really did see or sense the manifestation of their spirits.

Life after Life

Raymond A. Moody. Jr. M. D, in his book *Life after Life* recounts experiences reported by some of his patients and other people that went through a period where they seemed to be physically dead; but their bodies returned to independent physical life. When they recovered from their medical crises, they told Dr. Moody about what they believed were real experiences. (Called "Near Death Experiences," NDE) They believed they (their souls) had been to some place where "dead" people go. They said they learned that people who are not resuscitated remain in that realm and go on to another place that they could not return from. There were enough independent reports to make him believe that the experiences were real and not a "normal" consequence of being near death; and the patients had actually been in some other realm with dead people.

The sequence of events during the sojourn into this realm was usually the same for all people, even though none of the people reported all of the events. (This is typical of eyewitness testimony. It is rare that all witnesses to any event report exactly the same things.) Nevertheless, there was enough correlation for Dr. Moody to combine all the reports into a sequential list of what happened to those people during their "Life after Life" experiences. Dr. Moody states that none of the resuscitated patients were able to *completely* describe the unearthly experience. The list from Doctor Moody's book is condensed and slightly modified below.

Seeing their bodies: The person can see a body they usually recognize as their own. The body is in the condition that caused the "death." If death was caused by major damage to the body that is how they see it. Usually, the sight is not troubling.

Realizing they are dead: The person usually is aware of being dead because of the changed condition of existence, although sometimes the dead person hears people saying the person is dead. There is a feeling of floating or flying. Sometimes they can see and hear people working on the body; but they do not feel any of the things being done to the body. Some of them could see and hear people outside the room where the body was and understand what they were talking about.

Feelings of peace: A feeling of peace and comfort fills the consciousness. Sometimes people hear a strange noise. Some report the noise being unpleasant. Others hear beautiful music.

Going through a tunnel: Almost all people who have reported these near-death experiences say they went through a dark place: most of them describe it as a tunnel or a sewer. One person reported going through a valley that was compared with Psalms 23, *"The valley of the shadow of death."*

Seeing the light: Most of the people reported meeting an extremely bright light that seemed to have superior knowledge and intelligence. The Light "tells" the person to recall good and bad experiences. Christians identified the light as the spirit of Jesus Christ. Others did not assign any kind of identity to it.

They all agreed that the "voice" seemed to be present as or surrounded by the extremely bright light. Evidently, this examination is a self-judgment of the soul of the person who has just died clinically. Solomon must have had some knowledge of this. The book of Ecclesiastes ends with a passage describing the period of dying and the experience of bringing *"every work into judgment, with every secret thing, whether it be good, or whether it be evil."* It describes the judgment of the soul, implying that there will be consequences for failure to *"fear God and keep his commandments."*

Ecclesiastes 12:6-14 Or ever the silver cord be loosed, or the golden bowl be broken, or the pitcher be broken at the fountain, or the wheel broken at the cistern. Then shall the dust return to the Earth as it was: and the spirit shall return unto God who gave it...

...And further, by these, my son, be admonished: of making many books there is no end; and much study is a weariness of the flesh. Let us hear the conclusion of the whole matter: Fear God, and keep his commandments: for this is the whole duty of man. For **God shall bring every work into judgment, with every secret thing, whether it be good, or whether it be evil.**

Total recall: Stimulated by the Light, people can remember every thing they had done during physical life: all good things that were done, all bad things that were done, and all the things left undone that should have been done, as described in the Bible passage quoted above. Some people who had not been near-death, but seemed extremely close to dying have said, "My whole life flashed before me," which I assume is the same thing.

Meeting others: Some people reported meeting relatives, friends, or others who had died previously. Some reported meeting a stranger who guided them. Some people reported seeing "gray" spirits—evidently people who cannot give up physical life and are unable to advance in their new existence. Some of the gray spirits could be seen attempting to communicate with "living" people. The gray spirits seem to be unaware of the new person and other spirits that the "clinically-dead" person could see.

The limit: At some point in the experience, people are told they cannot remain in the new condition and must return to physical life, or they just start coming back without any comments.

Coming back: The person returns to the area of the body after leaving the "spiritual" environment. They might see what is happening around the physical body, as if they were floating above the area or they might go directly to their bodies. This is similar to what happened just after "dying." Then they "fall" into their physical bodies and have the pain and other feelings they had temporarily left behind. Many of them regret returning to life.

90 Minutes in Heaven: A Baptist minister, Don Piper, was killed in a terrible automobile accident and lay dead for 90 minutes; but he was revived with the help of a friend who felt he had to go to the crash scene before he knew his friend was in the crash. After reaching the scene, he felt compelled to keep praying for Don for more than an hour. He had begun singing hymns out loud when suddenly he heard the man who had been confirmed dead by paramedics singing along with him. The paramedics, who were beginning to leave the scene refused to believe that the victim was actually alive. Finally, they were sufficiently convinced that they returned to him and realized he was breathing; and he was rushed to hospitals that were equipped to treat such serious injuries. Don Piper was in terrible shape and almost died again several times during the first few months; but he survived and was able to walk reasonably well by the end of two years.

The Reverend Don Piper wrote a book with his friend, Cecil Murphy, *90 Minutes in Heaven*, in which he describes the time he was clinically dead for one and a half hours. He did not go through any of the steps mentioned by Dr. Moody, except being met by deceased relatives and friends and being returned to life. He reported being at the edge of heaven, which he described as well as he could. He wrote that it was wonderful and during the rest of his continuously painful life, many times he has regretted not staying there. However, he has discovered a special mission in helping many people to recover from serious accidents, encouraging people that were close to death, and comforting surviving relatives.

The following passage from the Bible reports the experience of "a man" that seemed to die and returned to life. This experience seems to be similar to what happened to Don Piper. Evidently, as indicated in the passage, his soul went into heaven or close to heaven after death but returned to physical life. Some scholars think the man mentioned was Paul, the writer of the passage, recounting something that happened to him after one of the times he almost died (permanently) because of being stoned or beaten for his preaching. It appears that he also went directly to heaven, similar to what Don Piper reported.

> **2Corinthians 12:1-4** It is not expedient for me doubtless to glory. I will come to visions and revelations of the Lord. I knew a man in Christ above fourteen years ago, (whether in the body, I cannot tell; or whether out of the body, I cannot tell: God knoweth;) such an one caught up to the third heaven. And I knew such a man, (whether in the body, or out of the body, I cannot tell: God knoweth;) How that he was caught up into paradise, and heard unspeakable words, which it is not lawful for a man to utter.

Lasting effects: People who have gone through this experience and remember it, usually feel that they should change how they were previously living. Many of them feel they should try to acquire more knowledge about everything and be more loving with others, especially family members.

The following passage, about the death and resurrection of a man called "Lazarus" may be referring to the person in the story about the rich man and the beggar. (Luke 16:19-26) If this is the same person, it is possible that the resurrected Lazarus told Jesus of something that he had experienced while his body was in the grave. This would qualify as one of Dr. Moody's life after life experiences, and helping confirm some of what he reported.

> **John 11:1** Now a certain man was sick, named Lazarus, of Bethany, the town of Mary and her sister Martha.

> **John 11:11-23** These things said he: and after that he saith unto them, Our friend Lazarus sleepeth; but I go, that I may awake him out of sleep.
>
> Then said his disciples, Lord, if he sleep, he shall do well. Howbeit Jesus spake of his death: but they thought that he had spoken of taking of rest in sleep.
>
> Then said Jesus unto them plainly, Lazarus is dead.
>
> Now Bethany was nigh unto Jerusalem, about fifteen furlongs off: And many of the Jews came to Martha and Mary, to comfort them concerning their brother. Then Martha, as soon as she heard that Jesus was coming, went and met him: but Mary sat still in the house.
>
> Then said Martha unto Jesus, Lord, if thou hadst been here, my brother had not died. But I know, that even now, whatsoever thou wilt ask of God, God will give it thee.
>
> Jesus saith unto her, Thy brother shall rise again.
>
> Martha saith unto him, I know that he shall rise again in the resurrection at the last day.

> **John 11:41-44** Then they took away the stone from the place where the dead was laid.
>
> And Jesus lifted up his eyes, and said, Father, I thank thee that thou hast heard me. And I knew that thou hearest me always: but because of the people which stand by I said it, that they may believe that thou hast sent me. And when he thus had spoken, he cried with a loud voice, Lazarus, come forth.
>
> And he that was dead came forth, **bound hand and foot with graveclothes:** and his face was bound about with a napkin. Jesus saith unto them, Loose him, and let him go.

Since Lazarus came forth *"bound hand and foot with graveclothes,"* he could not have walked out of the tomb. The passage does not mention how this happened; nor does it say whether he was in a vertical or horizontal position. Probably he was vertical. The passage probably would have mentioned it if he were floating out in a supine position. Gilbert R. Lavoie, M. D. in his book, *Unlocking the Secrets of the Shroud*, believes the body of Jesus was also raised into a vertical position when his body was resurrected. The Shroud probably is part of the *graveclothes* that were wrapped around his body before being entombed, after his crucifixion. An image of the crucified body of a man appears on the cloth: an image that could be the image of Jesus Christ as he was being lifted up from death. The marks on the image are consistent with what happened to Jesus before and during the crucifixion as related in the Gospels. The "shadows" on the body indicate it was lighted from above when the image was made, indicating that whatever caused the image to develop was either strong light, heat, or some other (spiritual?) kind of energy. This might be related to what happened when Jesus healed people: virtue (energy) was drawn from his own body to perform the "miracles" of healing.

Jesus brought another person back to the world of the living as recounted in the following passage.

> **Luke 7:11-15** And it came to pass the day after, that he went into a city called Nain; and many of his disciples went with him, and much people. Now when he came nigh to the gate of the city, behold, there was a dead man carried out, the only son of his mother, and she was a widow: and much people of the city was with her.
>
> And when the Lord saw her, he had compassion on her, and said unto her, Weep not. And he came and touched the bier: and they that bare him stood still. And he said, Young man, I say unto thee, Arise. **And he that was dead sat up, and began to speak.** And he delivered him to his mother.

These passages show that Jesus taught that the soul can be separated from the physical body and retain its personality.

Life after Death

The Bible, especially in the New Testament, clearly states some part of a human being survives after death of the physical body. Even though we have named it the *soul*, it does not mean that this part has been clearly described anywhere. We assume the soul is the part that we think is ourselves—our personalities. Most of us who believe in this kind of survival believe that when our souls depart from our bodies, we will continue to have memories of existence while on Earth; and once there, we will meet and recognize family members and friends we knew. Some people think that their souls will be transported directly to the end of time when they will be judged for what happened in their physical lives. Some think souls will rest quietly until the end of time. Some believe that souls will go directly to heaven if they were "good enough" or to hell if they were "not good enough." Some think that souls that were mostly good but not good enough to go to heaven go to a place of purification until it is purged of its sins and is pure enough to go on to heaven. Some people think that the soul enters a period of reminiscence and, when it is "ready," it enters the body of an unborn baby in "reincarnation."

Consequences of Behavior

No matter what they believe about the afterlife, most people believe that the things they do in this life will affect what happens after death of our physical bodies. In the following passages, eternal life is mentioned as an outcome of the way people live in this life. Evidently, not everyone will receive eternal life.

> **Luke 10:25-28** And, behold, a certain lawyer stood up, and tempted him, saying, Master, what shall I do to inherit eternal life?
> He said unto him, What is written in the law? how readest thou?
> And he answering said, Thou shalt love the Lord thy God with all thy heart, and with all thy soul, and with all thy strength, and with all thy mind; and thy neighbour as thyself.
> And he said unto him, Thou hast answered right: this do, and thou shalt live.

Conclusion

If a person actually is able to travel in some manner to a spiritual realm and remember the incident after returning to the material world, then the experience has demonstrated the ability to retain self-awareness, even after leaving the physical body. Nevertheless, life after life experiences do not prove that some part of a personality survives physical death; nor do spirit communications from the "dead;" but they provide powerful evidence that it does. If a personality can have experiences after being declared clinically dead that are similar to those many other people have had, the reports probably are true. Skeptics claim the brains of all people must go through a chemical change during "almost dead" states that causes their thoughts to proceed in this bizarre manner. Religious people are likely to accept the reports; but most of them place their own interpretations on them.

My wife, Betty, had an experience similar to those Dr. Moody discussed. She was suffering from a serious ailment that evidently took her into the "near-death" state. When she entered the hospital, she collapsed and went into a coma. Then she went through a tunnel and saw the bright light. She also saw her body and the people in the hallway as well as those in her room—before she woke up. What she "saw" happening in the vicinity of her body and in the hall outside her room was confirmed later. After the incident, she remained in serious condition for several days before she recovered her health. Betty thinks she was in a near-death condition and the experience was real.

I had an out of body experience during an operation. I could see my body on the operating table seemingly from a place above the table. I do not think I was near death; but I was unconscious, under an anesthetic. The most startling part of the experience was the sudden feeling of falling and landing on the recovery table. I did not have or else I do not remember any of the other experiences listed by Dr. Moody. Perhaps I was not "dead enough" to go through the entire sequence.

SPIRITUAL PERFECTION

In advancing toward spiritual perfection, you must recognize that you are not perfect, or in the words of Paul, you, and *all* other humans, have sinned and "*come short of the glory of God.*" (Romans 3:23) This verse means that in our present state of spiritual development we are not ready to enter the glory (presence) of God. Some Christians have taught that we can become perfect through accepting Jesus Christ as our savior. Some have taught that we can become perfect through accepting the presence of the Holy Ghost in our lives. Some believe that we can gradually become perfect by living a pure life. Some believe we can become perfect at the moment of death. Some religions teach that heaven can be reached by their members who convert or kill those who do not believe the way they do. Some believe we can become perfect after death in a spiritual place called Purgatory.

Roman Catholics believe Purgatory is not hell; but where souls exist in a spiritual world that seems to be material and they repent for sins committed during physical life, just as if they were in physical bodies. (If humans can work toward perfection in a protected spiritual realm, why does God expose humans to the dangers and temptations of physical life?)

The Bible has many references to perfection, or sanctification, which is sometimes defined as perfection. All Bible writers agree that perfection is necessary for seeing God—or entering into His presence. (Hebrews 5:5-10; Matthew 5:8, 48; Job 8:20; Ezekiel 28:12-17; 1John 4:16-21; Romans 6:13; 2Corinthians 4:6; Colossians 3:10; 1John 4:7; 1Corinthians 6:19; 1Corinthians 6:11; 2Thessalonians 2:13)

The Bible also has references to the inability of mankind to reach perfection while in the physical body. (Ecclesiastes 7:19-25; Proverbs 20:9; 1Kings 8:46; James 3:2; 1John 1:8; Romans 7:14-25; Philippians 3:12-14; 1Timothy 1:15; Psalms 19:12, 13, 51; Job 42:5, 6; Daniel 9:3-20)

Sister Teresa, called *The Saint of Calcutta*, well known for her life of sacrifice for the Lord's work, was one of the most saintly women that ever lived. She certainly worked diligently as a Christian missionary in India during most of her life. Yet, even she doubted her own faith; and she wondered about the existence of God and her soul. She wrote, "I have no faith—I dare not utter the words and thoughts that crowd in my head…" (*Come be My Light: The Private Writings of the Saint of Calcutta*) Perhaps only the most saintly people recognize their true weaknesses and their distance from perfection.

If perfection is unattainable while living in the physical world, what is the purpose of being born in a physical body? If this experience is a test of some kind, how can it be fair to humans that die while infants, while others live for many decades? The infant cannot even know what perfection is and cannot try to find it. This raises the question again, *what is the purpose of being born in a physical body?* If how we behave and believe in this life is the *final* test, the infant that dies as an infant cannot have lived long enough to "take the test." Many people believe that if children die before five, six, or seven, they will automatically go to Heaven. Some religions believe that only children who have been "properly" baptized will go to Heaven. Other children that die without being baptized go to some other, less wonderful place that is not hell, purgatory, or Heaven. (If either of these "rewards" is true, that is not fair to people who live into adulthood when they are exposed to multitudes of "tests.") If the soul is striving for perfection in some spiritual realm, where is that place? Why is it impossible to do that while in a physical body?

The Bible does not mention anything like Purgatory or give much information about where we go after death of the physical body. Jesus told a parable about a man, Lazarus, who had been dead but returned to life, when he reported that he had been in Heaven; but he could see a man in a place of torment. (Luke 16:20-31) Paul wrote of another man who had *"been out of the body,"* and saw the wonders of the afterlife and returned to tell part of the experience. (2Corinthians 12:2-4) The last two chapters of Revelation give us a brief description of *the holy city, new Jerusalem, coming down from God out of heaven,* which will be for *he that overcometh* (who) *shall inherit all things; and I will be his God, and he shall be my son.* (This is where) *the nations of them which are saved shall walk in the light.* However, this is not Heaven; because the holy city is coming down from Heaven. Nevertheless, the Holy City is a wonderful place. Evidently, it will be on a *new Earth*. There is no mention of what the saved, or perfected, souls will be doing there.

> **Revelation 21:1, 2** And I saw a new heaven and a **new earth**: for the first heaven and the first earth were passed away; and there was no more sea. And I John saw **the holy city, new Jerusalem, coming down from God out of heaven**, prepared as a bride adorned for her husband.

> **Revelation 21:10** And he carried me away in the spirit to a great and high mountain, and shewed me that great city, the holy Jerusalem, descending out of heaven from God,

> **Revelation 21:7, 8 He that overcometh shall inherit all things; and I will be his God**, and he shall be my son. But the fearful, and unbelieving, and the abominable, and murderers, and whoremongers, and sorcerers, and idolaters, and all liars, shall have their part in the lake which burneth with fire and brimstone: which is the second death.

> **Revelation 21:24** And **the nations of them which are saved shall walk in the light** of it: and the kings of the earth do bring their glory and honour into it.

Can We Ever Reach Spiritual Perfection?

If we ever expect to reach spiritual perfection, we must not do or think anything that separates us from God while in a physical body on Earth. The Bible contains many lists of various kinds of activities that are likely to cause separation from God. Idolatry, adultery, oppressing the poor and needy, violence, and robbery, are examples of "sinful" acts listed in Ezekiel 18:5-17. (Sin is anything that causes separation from God. In God's eyes, there are no minor sins.) Lawless and disobedient people, whoremongers, those who defile themselves with men, liars, and "*any other thing that is contrary to sound doctrine;*" are listed in 1Timothy. 1:9-10. The passage below states that those who follow the *works of the flesh* will not *inherit the kingdom of God*.

> **Galatians 5:19-21** Now the **works of the flesh** are manifest, which are these; Adultery, fornication, uncleanness, lasciviousness, Idolatry, witchcraft, hatred, variance, emulations, wrath, strife, seditions, heresies, envyings, murders, drunkenness, revellings, and such like: of

> the which I tell you before, as I have also told you in time past, that they which do such things shall not **inherit the kingdom of God**.

Admonitions in one of the oldest books in the Bible include many things that have been thought of as sins. One list is shown below.

> **Leviticus 19:13-19** Thou shalt not defraud thy neighbour, neither rob him: the wages of him that is hired shall not abide with thee all night until the morning. Thou shalt not curse the deaf, nor put a stumblingblock before the blind, but shalt fear thy God: I am the LORD. Ye shall do no unrighteousness in judgment: thou shalt not respect the person of the poor, nor honour the person of the mighty: but in righteousness shalt thou judge thy neighbour. Thou shalt not go up and down as a talebearer among thy people: neither shalt thou stand against the blood of thy neighbour: I am the LORD. Thou shalt not hate thy brother in thine heart: thou shalt in any wise rebuke thy neighbour, and not suffer sin upon him. Thou shalt not avenge, nor bear any grudge against the children of thy people, but thou shalt love thy neighbour as thyself: I am the LORD. Ye shall keep my statutes. Thou shalt not let thy cattle gender with a diverse kind: thou shalt not sow thy field with mingled seed: neither shall a garment mingled of linen and woollen come upon thee.

There are other lists and many other specific admonitions to do good and refrain from evil. The "Ten" Commandments, of course, are the basic rules set by God. Some others are less specific, such as *Follow righteousness, faith, charity, and peace, do not fear.* Jesus gave a message that summarizes in a few sentences what we should do while in the body.

> **Mark 12:28-33.** And one of the scribes came, and having heard them reasoning together, and perceiving that he had answered them well, asked him, Which is the first commandment of all? And Jesus answered him, The first of all the commandments is, Hear, O Israel; The Lord our God is one Lord: And **thou shalt love the Lord thy God with all thy heart, and with all thy soul, and with all thy mind, and with all thy strength:** this is the first commandment. And the second is like, namely this, **Thou shalt love thy neighbour as thyself.** There is none other commandment greater than these. And the scribe said unto him, Well, Master, thou hast said the truth: for there is one God; and there is none other but he: And to love him with all the **heart**, and with all the **understanding**, and with all the **soul**, and with all the **strength**, and to love his neighbour as himself, is more than all whole

burnt offerings and sacrifices. And when Jesus saw that he answered discreetly, he said unto him, Thou art not far from the kingdom of God. And no man after that durst ask him any question.

1John 4:16-21. And we have known and believed the love that God hath to us. **God is love**; and he that dwelleth in love dwelleth in God, and God in him. Herein is our love made perfect, that we may have boldness in the day of judgment: because as he is, so are we in this world. **There is no fear in love**; but perfect love casteth out fear: because fear hath torment. **He that feareth is not made perfect in love.** We love him, because he first loved us. If a man say, I love God, and hateth his brother, he is a liar: for he that loveth not his brother whom he hath seen, how can he love God whom he hath not seen? And this commandment have we from him, **That he who loveth God love his brother also.**

Why did God create humans with the tendency to commit sins? God must have known that Adam and Eve would violate the law about eating the fruit. It seems to me that humans live in a physical environment in order to grow and evolve to a higher spiritual level. I don't know why we were not made into Sons of God in the beginning—perhaps *what we are now is a continuation of a lengthy creation process,* lengthy by our standards. Jesus often used trees and other vegetation in parables—maybe we are like the seeds of trees that take a long time to begin to grow, reach maturity, and produce useful crops.

Life on Earth is *not a test* by God. It is an *opportunity* for us to learn how to *control our emotions, souls, minds, and bodies so we can get in harmony with God.* The ultimate objective of almost all religions throughout time has been harmony or union with God in Eternity. They all have different ideas what that means. The Bible accepted by Christians says we can become Sons of God meaning we can become partners with The Father, the Son, and the Holy Spirit. Christians believe Jesus of Nazareth, who was the only *begotten* Son of God, identified goals that would bring us to our objective of union with God; and he gave us instructions how each of us can determine the purpose why we were born on Earth. What Jesus said is not too much different from what most religions teach about good and evil, except they might have different definitions of what is good and what is evil.

SPIRITUAL GROWTH

Everlasting Life

We were created to grow spiritually, to evolve into something great—something actually capable of being with God, something eternal. This is our purpose in living on Earth.

> **John 3:16** For God so loved the world, that he gave his only begotten Son, that whosoever believeth in him should not perish, but have **everlasting life**.

Fundamental Christians believe that by accepting the divinity of Jesus, we can become adopted children of the Almighty God. I believe the words, "only begotten," in the King James Version mean that the physical body of Jesus of Nazareth contained the divinity of the only Son that was pre-existent and incarnated as a human being. Other versions of the Bible interpret this phrase meaning there is only *one* Son of God. They say, "One and only Son of God." This alternate interpretation implies that human beings cannot ever become Sons of God. Yet, other places in these versions indicate that we can reach that elevated state of spirituality. If we cannot become Sons of God, the meaning of creation, life, and death does not have the same significance for humans; our destiny is about at the level of Angels, which is not bad, but less significant than the other interpretation.

Reincarnation

If the soul can actually leave a human body, "travel" to some celestial place, and return to reside in that body again, as described in the preceding section, it is possible that such an experience could happen to a soul that has left a body that was no longer suitable. Then, after "residing" in a celestial place for a while, it could enter an unborn baby and start a new life.

Most Christian leaders and many Christian laymen do not accept the concept of reincarnation. They believe it is not compatible with the teachings of Jesus. It certainly is not compatible with what mainline religious leaders have taught. However, the criticisms that I have heard and read are against erroneous ideas about the process and the purpose of being reincarnated.

One criticism of the concept of reincarnation is that it weakens God's control over mankind. The idea is that it suggests that God is not powerful enough to reward and punish humans. The answer to that is that living in repeated lifetimes might be the way that God rewards and punishes.

Other popular arguments against reincarnation are based on the attitude toward any single lifetime. If people believe they will have multiple lives, they might think it will not matter what they do in any single lifetime. The belief that one is being punished in a current lifetime might cause some people to be fatalistic and fail to make efforts to improve themselves. Those who were born and live in pleasant conditions might believe they deserve what they have and become arrogant and oppressive of others who are not so fortunate. Some people who are cruel and harm or fail to help others might claim they are merely doing God's will because their victims deserve what they are getting. The argument against these comments is that many people behave and think like that without believing in reincarnation. In addition, we are told by God in the Bible to live every day as if it were the last day we will live.

Another popular complaint is that reincarnation makes the sacrifice of Jesus on the cross a foolish act that does not do anything for salvation of mankind. The argument against this is that part of our responsibility during any lifetime is to accept the divinity of Jesus Christ and realize that his sacrifice can move us farther along on our path toward our destiny—if we accept it and accept our share of responsibility for his torment and torture. The current belief about his sacrifice is that God demanded it in order to make amends for the sins of humans. This seems completely unfair to those who never get the opportunity to hear about it and accept it. That also means that it does not matter what we do. We do not have any responsibility or reward for our own acts, neither good nor bad; as long as we make the proper statement and perform certain rituals specified by various sects and cults. We can take a chance and wait until we are certain that we will die and then go through the required ceremony.

The criticisms might be against only the concept of reincarnation that the critics know of. There are several concepts accepted by large numbers of people, none of which is consistent with the teachings of Jesus.

The Hindu concept of reincarnation is that some part of humans continues to exist and be reborn after death of a physical body; but, that part might be reborn in any kind of living being, human, or lesser developed animals. The purpose of reincarnation is to purge the self of sin through karma, which is a law that states good deeds will be rewarded and bad deeds will be punished. Nevertheless, the person being punished or rewarded is unaware of previous actions and does not know why he or she is in a specific body and suffering with or enjoying his or her condition.

Plato taught that the soul existed in a celestial world and for some reason fell to Earth into a human body and will continue to be reborn in a physical body until the soul is purified through successive periods of learning. The purpose of being reincarnated is to make the soul pure so it can return to the celestial state.

Buddhism teaches that souls are being punished in successive incarnations because they cannot free themselves from earthly desires. Life in a physical body is punishment; and suffering is unavoidable. Rebirths will cease when the self is completely free of desire. At the end of its struggle, it will merge with the Supreme Being and lose its identity. Buddha taught that rebirth in a human body is rare; more often it is reborn in some other kind of animal.

The modern explanation, which is accepted in New Age philosophy, is not too different from Plato's idea but it does not describe the state of being that started the chain of rebirths.

Being reborn is a process of the soul continually progressing until it reaches the highest level of spiritual knowledge. The key to understanding the concept of reincarnation is *souls exist outside of the physical bodies,* whether in a spiritual place, or only in one or in many previous bodies. Several passages in the Bible indicate that God was aware of the existence of at least a few individual humans before they were born. Some Christian religions teach that these passages refer to what they call "predestination," a concept that fits the Hindu concept of karma.

Psalms 139:13-16. …thou hast covered me in my mother's womb. I will praise thee; for I am fearfully and wonderfully made; marvelous are thy works; and that my soul knows right well. **My substance was not hid from thee, when I was made in secret,** and curiously wrought in the lower parts of the earth. Thine eyes did see my substance, yet being imperfect; and in thy book all my members were written, which in continuance were fashioned, when as yet there were none of them.

Jeremiah 1:4-5. Then the word of the Lord came unto me, saying, **before I formed thee in the belly I knew thee;** and before thou came forth out of the womb I sanctified thee, and I ordained thee a prophet unto the nations.

Luke 1:31-32. …**thou shall conceive in thy womb and bring forth a son,** and shall call his name Jesus. He shall be great, and shall be called the son of the highest.

The Bible states that many humans were created at the same time. (Genesis 1:26-28) After creating humans, God rested. (Genesis 2:2) Therefore, it appears that this was the end of creation: no more humans were created; and all the souls that inhabit human bodies now were created at that time. The creation was a long time ago. Where were the souls during that time? The traditional answer might be that even though they were created many centuries ago by our time, God was able to bring them to our time instantly. If all the souls that have ever existed were not made during the sixth day of creation, then God did not rest; or if he resumed his creation because it was not perfect, why did he say "it is good" at the end of the sixth day. (Genesis 1:31) Quite a few passages in the Bible, both the Old and the New Testaments, can be interpreted as supporting the concept of reincarnation. However, most religions of the Western World interpret them in other ways. Evidently, different people will interpret according to their abilities to understand what God is telling us in the Bible.

Jesus did not teach everyone the same way. He deliberately spoke in parables that he explained later to his closest disciples. He told them what the parables actually meant, as demonstrated in Matthew 13:1-23, Mark 4:1-20, and Luke 8:4-15. This was the parable of the farmer casting seed on various kinds of ground. Jesus explained to the disciples that the seed was not something that would grow into a plant; but it was the word of God. Jesus meant that not everyone was capable of understanding his entire message. This passage probably gave support to the Gnostic belief that only some people were capable of understanding the mysteries of God's truths, and to Calvin's concept of predestination. It could also support the concept of reincarnation; some people are incapable of knowing certain things because their souls are at too low a level of spiritual development.

> **Matthew 13:10-16** And the disciples came, and said unto him, Why speakest thou unto them in parables?
>
> **He answered and said unto them, Because it is given unto you to know the mysteries of the kingdom of heaven, but to them it is not given.** For whosoever hath, to him shall be given, and he shall have more abundance: but whosoever hath not, from him shall be taken away even that he hath. Therefore speak I to them in parables: because they seeing see not; and hearing they hear not, neither do they understand.
>
> And in them is fulfilled the prophecy of Esaias, which saith, By hearing ye shall hear, and shall not understand; and seeing ye shall see, and shall not perceive: For this people's heart is waxed gross, and their ears are dull of hearing, and their eyes they have closed; lest at any time they should see with their eyes, and hear with their ears, and should understand with their heart, and should be converted, and I should heal them.
>
> But blessed are your eyes, for they see: and your ears, for they hear.

Paul repeated the explanation that only some people will understand God's true message.

> **Colossians 1:24-27** Who now rejoice in my sufferings for you, and fill up that which is behind of the afflictions of Christ in my flesh for his body's sake, which is the church: Whereof I am made a minister, according to the dispensation of God which is given to me for you, to fulfil the word of God; **Even the mystery which hath been hid from ages and from generations, but now is made manifest to his saints**: To whom God would make known what is the riches of the glory of this mystery among the Gentiles; which is Christ in you, the hope of glory:

I cannot believe that God created some people that were destined to never have the knowledge or ability to understand what was "*made manifest to his saints.*" I believe the term "saints" refers to people whose souls have advanced spiritually far enough to understand God's message through many reincarnations.

Definitions

The Soul is part of every living human being. It is eternal. It lives on after death of the physical body. The soul is immaterial, but controls a physical body during life of that body. In other words, Human Beings are souls that temporarily inhabit physical bodies. Most theists believe that the souls of all humans continue to exist after physical death in a spiritual realm; and they will be rewarded for doing good, or punished for evil deeds. Most Christians, Jews, Muslims, and Zoroastrians believe the soul can be reborn in either a new or a reconstructed previous body at the "final judgment" in what is called *resurrection* It retains its identity in physical life and in the spiritual world. My dictionary says soul is the immaterial essence or the animating force that gives a body life. The same dictionary says that spirit is the animating principle that gives life to physical beings. The Bible, I think, is a little clearer, at least by inference. Matthew 16:26 shows that the soul is a significant part of the human body and not just an animating force; for it states that a man can lose his own soul or exchange it for something else just as if it were a physical commodity that could be bartered. If the word soul merely indicated the animating force, then there would not be any sense in exchanging the animating force (physical life) for something else. In other words, the person trading his soul would simply die physically.

> **Matthew 16:26** For what is a man profited, if he shall gain the whole world, and lose his own soul? or **what shall a man give in exchange for his soul**?

The Spirit is the driving force of living bodies. It is frequently used to refer to the soul when it is separated from the body, also called an apparition or a ghost. The following passages from the Bible indicate that the word spirit refers to a something that might be visible, or it might be felt; but it is an immaterial representation of a person or a spiritual presence. I have read about claims that there is a slight weight loss just as a person dies, which indicates that some material or vapor leaves the body when a person dies. Supposedly, this was done with sensitive scales below the beds of people that were expected to die. (I do not know if this was scientifically conducted or if the story is even true. It is repeated here to suggest the need for more, and better scientific, study about the subject of death.)

> **Luke 24:36, 37** And as they thus spake, Jesus himself stood in the midst of them, and saith unto them, Peace be unto you. But they were terrified and affrighted, and supposed that **they had seen a spirit**.

> **Job 4:15-27** Then **a spirit passed before my** face; the hair of my flesh stood up: It stood still, but I could not discern the form thereof: an image was before mine eyes, there was silence, and I heard a voice, saying, Shall mortal man be more just than God? shall a man be more pure than his maker?

Resurrection means bodies of humans who have died will be resurrected at some time in the future after an important series of events take place. The concept seems to have originated in ancient Egypt. The ancient Egyptians embalmed their dead so that their bodies would be available to them when the grand resurrection takes place. Many Jews, Christians, and Muslims continue to believe this will happen; but the resurrected bodies will not be dried corpses; they will be completely reconstructed bodies that will be similar to their original bodies.

Jesus continued teaching about the resurrection. He altered the concept that the Sadducees had about that condition. He told them resurrected bodies would not be ordinary humans; but would be like angels. Nevertheless, he did support the idea of resurrection.

> Mark 12:18-25 Then come unto him the Sadducees, which say there is no resurrection; and they asked him, saying,
>
> Master, Moses wrote unto us, If a man's brother die, and leave his wife behind him, and leave no children, that his brother should take his wife, and raise up seed unto his brother.
>
> Now there were seven brethren: and the first took a wife, and dying left no seed. And the second took her, and died, neither left he any seed: and the third likewise. And the seven had her, and left no seed: last of all the woman died also.
>
> In the resurrection therefore, when they shall rise, whose wife shall she be of them? for the seven had her to wife.
>
> And Jesus answering said unto them, Do ye not therefore err, because ye know not the scriptures, neither the power of God? For

when they shall rise from the dead, they neither marry, nor are given in marriage; but are as the angels which are in heaven.

Judaism Concept of Death is that it is part of the cycle of life and is not a tragedy except how it affects the family and close friends. Most Jews believe in an afterlife; but the holy writings do not include much discussion of it. One of the statements of faith developed by Moses Maimonides declares the dead will be resurrected. The concept of physical resurrection evidently was widely accepted by Jews of Jesus' time. It was mentioned casually at the tomb of Lazarus before Jesus raised him from the dead.

> **John 11:23-24** Jesus saith unto her, Thy brother shall rise again. Martha saith unto him, I know that he shall rise again in **the resurrection at the last day.**"

Most Christian Churches teach resurrection as an article of faith. The Roman Catholic Church states it is a doctrine of revealed truth. The Apostle's Creed states, "I believe in the Holy Ghost, the holy Catholic Church, the communion of saints, the forgiveness of sins, **the resurrection of the body**, and the life everlasting." Since Christians believe Jesus of Nazareth was crucified, died, was buried, and rose from death, they find it easy to believe they, too, will be resurrected at some time.

Muslims also teach the resurrection. The 75th chapter of the Koran is titled, "The Chapter of the Resurrection." It starts out, "*I do call to witness The Resurrection Day.*" In this chapter, the subject of resurrection is discussed from the viewpoint of mankind. One of the five pillars of faith is "The dead will be resurrected on the last day, judged by Allah, and sent to Heaven or Hell."

Incarnation refers to the act of giving life to a physical being by the insertion of a soul into a physical body. One of the dictionary definitions is it is, "The embodiment of a deity or a spirit in an earthly form." Many Christians believe that this occurs very soon after conception takes place. Many other people either believe a baby does not have a soul until after it leaves the body of its mother, or they do not believe in the concept of an eternal soul.

Reincarnation means that a soul that existed in a physical body at some time in the past and left that body returns to exist in a completely new physical body. Reincarnation is the rebirth of a soul, usually in a new human body. Transmigration can mean the same thing; but some of those using this term include the possibility of rebirth in bodies of other kinds of beings.

Many people around the world accept the general concept. The Hindu sects are the most well known groups that teach the concept. Buddhists do not believe that rewards and punishments are meted out in a hereafter; suffering occurs in the physical body; and the only reward is to cease being reincarnated so that the soul can join God.

Many people involved in the new religions accept the concept: New Age followers, Spiritualists, Rosicrucians, Wiccans, a few Christian groups that developed new teachings, followers of Edgar Cayce, and other not so well known groups.

Retribution: Even though a Supreme Being or a spiritual or secular leaders might forgive deviations from the proper code of conduct or disobeying restrictions, usually there are consequences for failures to adhere to religious rules. The Roman Catholic Church teaches that retribution occurs in penance while in the body and/or for a period in Purgatory after death for those not evil enough to be sent to eternal punishment. Some Jews also believe that souls will spend time in a place like purgatory. In religions that believe in reincarnation, retribution occurs in subsequent lives. Hindus believe in *Karma,* the name for the process of being rewarded or punished in one life for what happened in previous lives. Most Christian religions teach there will be no serious retribution or other consequences for failures if sins are confessed and repented and transgressors demonstrate repentance or perform some act or ritual that is approved by the leaders of the religion.

Which is more reasonable: The belief that human bodies will be reconstructed in pretty much the way they were in one physical life; or the belief that human souls will be reborn in new physical bodies that will be conceived and be born in the normal manner? In the resurrection theory, the soul is tormented or rewarded in a spiritual realm. In the reincarnation theory, the soul suffers and is rewarded in a physical body.

Reincarnation in History

Many people in the past believed in reincarnation. Some of the Egyptian religions might have included belief in reincarnation. They were aware of the cycles of nature in the flooding of the Nile and in rebirth of vegetation and might have reasoned that human souls might go through the same process. The Hindu religion taught the concept thousands of years ago. Like the Egyptians, they recognized the cycles of nature. They believed that the entire Universe goes through endless cycles of creation, material existence, and destruction. The Mayans of Central America also were aware of the cyclical nature of the Universe and they probably accepted the concept of reincarnation in some form. They believed, like the Hindus, that the entire Universe exists for a while, is destroyed, and then is reconstructed.

Pythagoras and Plato, philosophers in ancient Greece, held to the idea that souls go through many rebirths. The Druids, a religion of ancient Britain, believed in reincarnation. The early Jews did not teach reincarnation. It was not until the development of Kabbalism that some Jews began to consider the idea. Some of the early groups that claimed to be Christians, such as the Gnostics and the Carthari, accepted the concept; but there is no proof that any of the independent groups that finally gathered as the ancestor of modern Christianity ever taught it. As far as can be discovered, the early leaders of Christian groups always rejected the concept. Some modern Christians believe Origen (185?-254?), one of the earliest Christian scholars, taught reincarnation. However, he clearly rejected the doctrine during the last years of his life. The Council of Constantinople, in 533AD, declared teaching belief in reincarnation was a heresy. Nevertheless, many modern Christians believe that Jesus Christ at least tacitly accepted the concept.

Gnosticism, Arianism, and Catharism taught reincarnation of souls. Partly because of this teaching, the Roman Catholic Church classed them as heretics and attacked them to extinction. This removed and ended teaching of the concept of reincarnation in any major part or denomination of Christianity.

Islam: Mainstream Islam rejects the concept of reincarnation. Believing in reincarnation in such a way that it could be interpreted as a denial of resurrection may constitute apostasy in Islam. Even so, a very few Muslim groups believe in reincarnation, claiming that this concept is mentioned in Koran 2.28. Mainstream Islam rejects this understanding of the verse, claiming that it refers to the worldly human life and the consequent resurrection in the future. However, Koran 3:27 may also indicate support for the theory of reincarnation.

> **Koran 2.28** How do you deny Allah and you were dead and He gave you life? Again He will cause you to die and again bring you to life, then you shall be brought back to Him.

> **Koran 3:27** Thou [God] makest the night to pass into the day and Thou makest the day to pass into the night, and **Thou bringest forth the living from the dead** and Thou bringest forth the dead from the living, and Thou givest sustenance to whom Thou pleasest without measure.

Judaism and Kabbala: Some Jews, especially Orthodox and Hasidic Jews, believe souls might be reincarnated. Among those who accept the concept of reincarnation, some believe all souls go through the cycle of life and death many times; others believe it only happens in special circumstances, such as when a soul must complete some kind of activity. However, they believe souls of Jews are reborn into Jewish families, unless more Jewish souls seek rebirth than there are Jewish families ready for them, such as occurred after the Nazi holocaust. Some Jews believe only the righteous are reincarnated in the process of *tikkun olam,* meaning "mending the world." A group of Jewish scholars has been examining alleged claims by Jews and Gentiles they are reincarnated Jewish souls.

A classic work of the Kabbala, *Shaar ha Gilgulim* ("Gate of Reincarnations") of Isaac Luria (1534-1572 AD), describes complex laws of reincarnation of the soul. Hebrew scholars do not believe reincarnation is mentioned in the Hebrew Bible or in the classical rabbinic works, Midrash, Mishna, and Talmud. Nevertheless, during the late thirteenth century, many in the Jewish community began to accept the concept. Jews who accepted this idea accepted it because it solved many of the apparent conflicts of a good God and the suffering of good people.

Critics of the concept worried that if people believed in reincarnation, they might not accept many of the rules and laws established to guide people through a single lifetime. In addition, the concept of the belief in resurrection of the physical body raised the question of which physical body would be resurrected since there might have been very many previous bodies.

The Bible and Reincarnation

Without realizing it, people who believe Jesus is the Son of God, tacitly accept the concept of reincarnation. Most Christians believe that Jesus of Nazareth existed long before Abraham lived. None of them says that the physical body of Jesus existed before Abraham—it was the soul or spirit within the body that existed previously. They say this was a special case. Nevertheless, Christians accept this one case as proof that a preexistent soul can enter the body of a baby.

The following verses from the gospel of John state that the personality of the man Jesus existed before Abraham. Whether this is an allegory, a special case, or something else, John and others that heard Jesus evidently accepted the concept literally. Most modern Christians accept this as meaning Jesus was part of God at creation, without trying to explain how the immortal Jesus became the physical Jesus. It is not clear what his enemies of his day believed, but they wanted to stone him for saying such a thing. Perhaps the most controversial statement was when Jesus said, "I am." That is the meaning of "Jehovah/Yahweh." By saying that, Jesus was claiming to be part of God, further supporting the concept of pre-existence.

> **John 8:56 58** Your father Abraham rejoiced to see my day: and he saw it, and was glad. Then said the Jews unto him, Thou art not yet fifty years old, and hast thou seen Abraham? Jesus said unto them, Verily, verily, I say unto you, **Before Abraham was, I am**.

Another section of John's gospel tells of a man blind from birth. Evidently, the disciples believed the man had some kind of opportunity for sinning before being born, because they asked Jesus if the man had sinned to cause being born with that handicap. Evidently, the disciples must have believed that he sinned either while in the womb, or in a previous existence, Since Jews do not object to abortions at any state of growth, evidently they do not think an unborn baby has a soul and is incapable of doing good or bad things. Therefore, the disciples must have believed the man sinned in some previous existence, either on Earth or in some spiritual realm.

> **John 9:1-3** And as Jesus passed by, he saw a man which was blind from his birth. And his disciples asked him, saying, Master, **who did sin, this man, or his parents, that he was born blind?** Jesus answered, Neither hath this man sinned, nor his parents: but that **the works of God should be made manifest in him.**

I have heard preachers interpret this passage to mean that God made the child that grew into the adult "*blind from his birth*" just so Jesus could heal him. I have a difficult time believing that. I do not think God would punish the man and his family for many years just so Jesus could create a miracle when the man was grown. It seems more likely that the man was born blind in order to teach *him* or *his parents* some spiritual message. According to the above passage from the Bible, rebirth, whatever the conditions of physical life might be, is an opportunity for spiritual growth—"*that the works of God should be manifest in him.*" We are expected to learn how to become Sons of God.

The Bible has other passages that imply reincarnation. In Jeremiah, the writer is told that God knew him before he was born. Malachi predicted that God will send *Elijah the prophet* at some time in the future, giving rise to the belief by some people that either Jesus or John the Baptist was the reincarnation of this person. (Matthew 17:10-13) Jesus told the disciples that Elijah (Elias) had returned as John the Baptist.

> **Jeremiah 1:4, 5** Then the word of the LORD came unto me, saying, **Before I formed thee in the belly I knew thee;** and before thou camest forth out of the womb I sanctified thee, and I ordained thee a prophet unto the nations.

> **Malachi 4:5, 6** Behold, **I will send you Elijah the prophet** before the coming of the great and dreadful day of the LORD: And he shall turn the heart of the fathers to the children, and the heart of the children to their fathers, lest I come and smite the Earth with a curse.

> **Matthew 17:10-13** And his disciples asked him, saying, Why then say

> the scribes that Elias must first come? And Jesus answered and said unto them, Elias truly shall first come, and restore all things. But I say unto you, That **Elias (Elijah) is come already,** and they knew him not, but have done unto him whatsoever they listed. Likewise shall also the Son of man suffer of them. Then the disciples understood that he spake unto them of John the Baptist.

Luke 1:17 seems to indicate that John was not the reincarnated Elijah; but he was merely acting in the "*spirit and power of Elias (Elijah).*" However, Malachi (above) clearly states that God will "*send you Elijah the prophet.*"

> **Luke 1:17** And he shall go before him in the **spirit and power of Elias**, to turn the hearts of the fathers to the children, and the disobedient to the wisdom of the just; to make ready a people prepared for the Lord.

Evidently, at least some of the Jews of Jesus' day believed in reincarnation for they suspected that Jesus was John the Baptist, Elijah, Jeremiah, or some other prophet reborn, as indicated in the following passage. (If Jesus were John the Baptist, that would have been spirit possession, not reincarnation; because Jesus and John were born only a few months apart and Jesus was still alive when John was executed.)

> **Matthew 16:13-16** When Jesus came into the coasts of Caesarea Philippi, he asked his disciples, saying, Whom do men say that I the Son of man am? And they said, **Some say that thou art John the Baptist: some, Elias; and others, Jeremias, or one of the prophets.** He saith unto them, But whom say ye that I am? And Simon Peter answered and said, Thou art the Christ, the Son of the living God.

A verse in the book of Job states that *man is born unto trouble*. Job also says that trouble does not come by itself, implying that men are responsible for their trouble because of something happening before being born in a body.

> **Job 5:6,7** Although affliction cometh not forth of the dust, neither doth trouble spring out of the ground; Yet **man is born unto trouble**, as the sparks fly upward.

The entire book of Job implies reincarnation and the operation of karma, the Hindu explanation for inequalities of life. Bible scholars are inclined to classify this book as allegorical; if it is; then it could be an allegory about karma. The deaths of Job's family and servants would have been unfair and cruel to them. I find it difficult to believe that a loving God would allow that to happen. I think the test of Job took place in another incarnation.

After Job had passed his test, the Lord took assets away from Job's accuser, and gave many more assets to Job. This punishment probably occurred in a subsequent incarnation. Even though Job lived "*a hundred and forty years*," it is difficult to understand that he could have recovered all the things that God gave to him in the years after his test was over. (Job 42:7-17)

There is a belief among many Christians (Calvinists) that some people are predestined to certain conditions; in other words, only some people are capable of being saved and going to heaven. Paul wrote about this in his letters. In Romans, Paul indicates that God was acquainted with in some way (foreknew) some people before being born on Earth. Paul says, "*For whom he did foreknow, he also did predestinate...*" (Romans 8:28-30) If this is so, God either intentionally creates some people in either a good or a deplorable condition, or the conditions of people result from their own actions at some prior time. Therefore, those who accept the concept of predestination must believe that the souls of humans existed some "place" at some "time" prior to being born. Calvinists explain the concept of God's foreknowledge means that God could foresee how the person would behave after being born. If some people are destined for damnation, why does God allow such souls to be born. It would be more merciful to prevent the soul from being born into a physical body that was destined for eternal punishment.

In Matthew 19:12 Jesus states that some men are born without the normal male capabilities without stating the reason why this happens. Perhaps Jesus meant to imply reincarnation.

Ephesians 1:5 Having **predestinated us** unto the adoption of children by Jesus Christ to himself, according to the good pleasure of his will,

Romans 8:28-30 And we know that all things work together for good to them that love God, to them who are the called according to his purpose. For **whom he did foreknow, he also did predestinate** to be conformed to the image of his Son, that he might be the firstborn among many brethren. Moreover whom he did predestinate, them he also called: and whom he called, them he also justified: and whom he justified, them he also glorified.

Matthew 19:9-12 And I say unto you, Whosoever shall put away his wife, except it be for fornication, and shall marry another, committeth adultery: and whoso marrieth her which is put away doth commit adultery. His disciples say unto him, If the case of the man be so with his wife, it is not good to marry. But he said unto them, All men cannot receive this saying, save they to whom it is given. **For there are some eunuchs, which were so born from their mother's womb:** and there are some eunuchs, which were made eunuchs of men: and there be eunuchs, which have made themselves eunuchs for the kingdom of heaven's sake. He that is able to receive it, let him receive it.

The following passage introduces a term that causes confusion and conflict within and outside of the Christian community. Jesus told Nicodemus that he must be "born again." The commonly accepted interpretation of this term among Christians is that this is allegorical, describing a spiritual experience. Christians believe this is a new experience, unique to Christianity, yet in verse 10 (shown below) Jesus seemed to think that Nicodemus should already understand what it means, implying that the concept was well-known among "*...master(s) of Israel...*" Jesus did not make the meaning clear, so we now have the problem of understanding the term based on our interpretations of it and other material in the Bible. Being "born again" might mean being reincarnated.

John 3:1-15 There was a man of the Pharisees, named Nicodemus, a ruler of the Jews: The same came to Jesus by night, and said unto him, Rabbi, we know that thou art a teacher come from God: for no man can do these miracles that thou doest, except God be with him.

Jesus answered and said unto him, Verily, verily, I say unto thee, **Except a man be born again, he cannot see the kingdom of God**.

Nicodemus saith unto him, How can a man be born when he is old? can he enter the second time into his mother's womb, and be born?

Jesus answered, Verily, verily, I say unto thee, Except a man be born of water and of the Spirit, he cannot enter into the kingdom of God. That which is born of the flesh is flesh; and that which is born of the Spirit is spirit. Marvel not that I said unto thee, Ye must be born again. The wind bloweth where it listeth, and thou hearest the sound thereof, but canst not tell whence it cometh, and whither it goeth: so is every one that is born of the Spirit.

> Nicodemus answered and said unto him, **How can these things be?**
>
> **Jesus answered and said unto him, Art thou a master of Israel, and knowest not these things?** Verily, verily, I say unto thee, We speak that we do know, and testify that we have seen; and ye receive not our witness.
>
> If I have told you earthly things, and ye believe not, how shall ye believe, if I tell you of heavenly things? And **no man hath ascended up to heaven, but he that came down from heaven,** even the Son of man which is in heaven. And as Moses lifted up the serpent in the wilderness, even so must the Son of man be lifted up: That whosoever believeth in him should not perish, but have eternal life. For God so loved the world, that he gave his only begotten Son, that whosoever believeth in him should not perish, but have everlasting life.

The verse that says, *"...no man hath ascended up to heaven, but he that came down from heaven,"* indicates that in order to reach heaven, a soul must come down from heaven. If Jesus was the only one that could reach this goal, then the entire concept of Christianity is meaningless. However, Jesus said in the last verse quoted above that whoever believed in the only begotten Son, would have everlasting life. This clearly implies that *anyone* can achieve this goal. This entire passage indicates it is necessary to be born on Earth after coming from at least one previous existence in heaven, in order to reach it again.

A peculiar condition is described in the following passage from Luke indicating there are spiritual entities that can take over the bodies of some people. Jesus cured people of what appears to have been either epilepsy or some form of mental disability. In the Bible, the condition is blamed on control of the person's body by a being from the spiritual world. These entities are called "demons" or "unclean spirits," meaning they are evil. A modern term for this condition is "possession" by a foreign spirit existing in a body while the birth-soul still lives in that body. Medical personnel prefer to explain the actions described in the Bible as "epilepsy," "bi-polar disorder," or something more professional. However, Jesus cured the victims with a few words instead of lengthy treatments or expensive medications, indicating that perhaps the demon explanation is simpler and the ailment easier to cure—if the proper technique is used. The Roman Catholic Church employs priests, called exorcists, who have been trained to remove "unclean" spirits from human beings.

> **Luke 9:38-42** And, behold, a man of the company cried out, saying, Master, I beseech thee, look upon my son: for he is mine only child. And, lo, **a spirit taketh him**, and he suddenly crieth out; and it teareth him that he foameth again, and bruising him hardly departeth from him. And I besought thy disciples to cast him out; and they could not. And Jesus answering said, O faithless and perverse generation, how long shall I be with you, and suffer you? Bring thy son hither. And as he was yet a coming, the devil threw him down, and tare him. **And Jesus rebuked the unclean spirit, and healed the child,** and delivered him again to his father.

A 1957 movie, *Three Faces of Eve*, with Joanne Woodard, described a woman who had three different personalities. A psychiatrist diagnosed her as schizophrenic and cured her with a lengthy treatment after he discovered she had suffered serious emotional trauma in childhood, which was the cause of her inability to control her body or her emotions. What happened to Eve could have been possession by at least one foreign spirit. If spirits can take over the control of adult physical bodies, then the basic concept of reincarnation uses the same process in a more orderly manner. Both reincarnation of a previously living soul into an unborn child and the imposition of a pre-existing soul or spirit on a mature physical body involve what I suspect is the same kind of spiritual mechanism. (Possession takes place when some kind of trauma weakens a person's ability to retain his or her personality and characteristics.)

Proof of Experiences Involving Reincarnation

Under hypnosis, some people apparently have been regressed to a previous existence, supposedly indicating reincarnation. Some of these reports have been partially verified by checking on information about the previous life given by hypnotized subjects. However, other possible explanations have been made for some of the reported confirmations. In addition, statements made under hypnosis are unreliable as evidence. There is too much opportunity for fraud—either intentional or unintended. Therefore, hypnotic regressions are not proof that anyone, much less everyone, has been reincarnated, although some cases have provided pretty good evidence of reincarnations.

Some Eskimos routinely test a newborn child for indications it is the reincarnation of a recently deceased relative. They offer the newborn an assortment of objects, some of which were possessions of a dead relative. Some children have correctly chosen objects previously owned by the relative. Of course, the correct choices might merely be coincidences, or the chosen object might have merely been the most attractive to a child.

Hindus believe in reincarnation as part of their religion. In India, a government agency investigates reported cases of children claiming knowledge of unrelated dead people who had lived far from where the children were born and lived. Many of the investigations uncovered evidence that verified the possibility of reincarnation. Again, just because a relatively few children's memories checked out, does not prove that they or any other people were reincarnated: what seem to have been accurate memories could have been the result of telepathic communications. However, as the children grew, most of them lost memories of things, places, and events that they "recalled" as children, which might indicate that many of us have had memories of previous lives but they were forgotten before we got old enough to realize what they represented.

Child prodigies, who have amazing skills that usually are acquired only after a lifetime of study and practice, might be using skills acquired in previous lives. Just as in the reported cases of children "recalling" things from previous lives, child prodigies also often lose their superior skills as they grow more mature. Perhaps when the mind acquires more-and-more new experiences, thoughts from previous lives get "crowded" out.

Very few members of organized religions of the Western world profess belief in reincarnation, although in private conversations, I have found that there are quite a few people of European descent that do not completely dismiss the possibility of such a thing happening. I once conducted an informal, unscientific poll with 18 fundamental Christians. Only half of the group said they did not believe in reincarnation. The other half made statements from, "absolutely yes," to "maybe," with about one quarter of the total group saying that they believed in it.

An interesting thing about this poll illustrates the reluctance of Christians to admit belief in reincarnation. The night before I took the poll, I was having dinner with the pastor of our church. We were discussing the many varied beliefs among Christians in various congregations. I claimed most people do not admit belief in something that is not popularly accepted—a common problem in conducting any poll. Since we both had separate Sunday school groups with people about the same ages, we agreed to poll our groups to see if any of them believed in the concept of reincarnation. I gave my group opportunity to write their answers on paper, which was folded for secrecy, with the results described above. Later, the pastor told me no one in his group admitted they accepted the concept in any degree. As the discussion continued, I learned that he had asked his group to raise their hands if they believed in reincarnation. I told him that proved my point; but he changed the subject.

Personal Experience: When I was about five or six, I heard someone talking about Beverly Hills. That immediately attracted my attention and I asked if we could go there. When I learned that it was a park a few miles away in the heart of Chicago, Illinois, I lost interest in it. A year or so later, I was in an automobile that passed a sign that said "California." I asked my dad if we could go there immediately. Again, I lost interest in it when I learned that it was a diagonal street in Chicago. For some reason, I always seemed to know about things in the state of California, especially the Los Angeles area. After I graduated from Kansas University, I was with a group of family members that drove to California, and I felt relieved that I was going there. I felt at home, especially in the Hollywood area. A couple years later, I took the first job that was offered to me in the Los Angeles area. The first few years I lived there, I enjoyed driving around the western part of the city. Once, when I passed a short street in Beverly Hills, I felt it was familiar. It seemed that I had lived there at one time. I do not claim that I was reincarnated from someone who had lived there; but I wonder from where these thoughts or "memories" originated.

Spirit and Body

In the following passages, the Apostles seem to have believed that the spirit of a living person could move and be seen without the physical body being present.

> **Matthew 14:24-17** But the ship was now in the midst of the sea, tossed with waves: for the wind was contrary. And in the fourth watch of the night Jesus went unto them, walking on the sea. And when the disciples saw him walking on the sea, they were troubled, saying, **It is**

a spirit; and they cried out for fear. But straightway Jesus spake unto them, saying, Be of good cheer; it is I; be not afraid.

In Acts 12:15, a Jewish woman, who knew that Peter had been put in jail, did not immediately realize that he was really at the door when she saw him. She thought it was "*his angel*." Evidently, she believed it was possible for some part of a person to travel outside of the body.

Acts 12:13-15 And as Peter knocked at the door of the gate, a damsel came to hearken, named Rhoda. And when she knew Peter's voice, she opened not the gate for gladness, but ran in, and told how Peter stood before the gate. And they said unto her, Thou art mad. But she constantly affirmed that it was even so. **Then said they, It is his angel**.

Christian scholars do not agree when the spirit of the Son of God entered the body of Jesus of Nazareth. Some think this spirit was with the physical Jesus from conception. Some think this spirit entered the physical body of the man Jesus at the time John the Baptist saw *"the Spirit descending from heaven like a dove,"* and remained on him. Neither of these beliefs challenges the purity, dedication, or love of the person of Jesus. His birth was a miracle and his entire life was completely pure and sinless. At the time of his baptism, Jesus was thirty years old. In the time of Jesus, people would not accept the words of a person younger than thirty years old, when men were considered finally grown-up. Therefore, there would have been no reason for the divine spirit of the Son of God to be in a physical body that was not fully matured.

John 1:29-34 The next day John seeth Jesus coming unto him, and saith, Behold the Lamb of God, which taketh away the sin of the world. This is he of whom I said, After me cometh a man which is preferred before me: for he was before me. And I knew him not: but that he should be made manifest to Israel, therefore am I come baptizing with water.

And John bare record, saying, I **saw the Spirit descending from heaven like a dove,** and it abode upon him. And I knew him not: but he that sent me to baptize with water, the same said unto me, Upon whom thou shalt see the Spirit descending, and remaining on him, the same is he which baptizeth with the Holy Ghost. And I saw, and bare record that this is the Son of God.

Jesus frequently used the term "son of man," indicating some entity or spirit separate from the being of the Son of God. In Matthew 17, Jesus said, "*Likewise shall also the Son of man suffer of them.*" Jesus did not say the Son of God would suffer, indicating that only the physical part would suffer. At the moment of his death, Jesus cried out, asking why God had forsaken him. If the spirit of the Son of God had been with Jesus during all or part of his life, it had to depart from the body or the body could not have died.

During the final moments of the crucifixion, the body of Jesus continued to live for a little while without the Spirit of God, but died shortly afterwards. Later, in the tomb, after physical death, the spirit returned to the body and the body was resurrected. However, the resurrected body evidently was not the same as it had been and was capable of performing tasks not previously demonstrated.

> **Matthew 27:46-50** And about the ninth hour Jesus cried with a loud voice, saying, Eli, Eli, lama sabachthani? that is to say, **My God, my God, why hast thou forsaken me?** Some of them that stood there, when they heard that, said, This man calleth for Elias. And straightway one of them ran, and took a spunge, and filled it with vinegar, and put it on a reed, and gave him to drink. The rest said, Let be, let us see whether Elias will come to save him. Jesus, when he had cried again with a loud voice, yielded up the ghost.

Becoming Sons of God

The key phrase in Christianity is "Son of God." This is what religion, science, and physical evolution is all about—spiritual evolution toward becoming Sons of God. Some Christian religions teach that all members of their faith become Sons of God or equivalent by accepting their doctrines and performing the correct rituals. I think that is a little presumptuous.

> **Philippians 2:15** That ye may be blameless and harmless, **the sons of God,** without rebuke, in the midst of a crooked and perverse nation, among whom ye shine as lights in the world;

Life on Earth is not a *test* by God; nor is it a place of punishment and suffering for previously committed sins. It is an *opportunity* for us to learn how to control our bodies, minds, and spirits and get in harmony with God. The ultimate objective of almost all religions throughout time has been harmony or union with God in Eternity. They all have different ideas what that means. They all have different ideas about God. The Bible accepted by Christians says we can become Sons of God meaning we can become *partners* with the Father, the Son, and the Holy Spirit. Christians believe Jesus of Nazareth, who was the only *begotten* Son of God, identified goals that would bring us to our objective of union with God, and gave us instructions how each of us can determine the purpose why we were born on Earth.

Luke 20:34-36 tells us that those who get to heaven will not marry or be given in marriage and will be as the Angels. Even so, the sons of God, those who are worthy of the term, whether male or female in this world will be greater than the angels. 1Corinthians 6:3 says, "*...we shall judge angels...*" In the KJV Luke 20:36, Jesus is quoted using the term, "children of God," so He must have meant that both sexes would be in heaven. Evidently, the gender differences apparent on Earth will not continue in Heaven, even though the Bible indicates that we will recognize our family members, meaning physical and sex differences will be apparent. However, all those who reach that level will have the authority and responsibilities of Sons of God.

Continued Spiritual Growth

In the discussion of physical evolution, it was pointed out that death of prior specimens of an evolving being had to die so there would be room and resources for the physical beings that had evolved to higher levels. The same need for physical death is required in order to continue evolution of our souls. Our souls must grow spiritually. Based on what religions of the world have taught us, this growth occurs while in physical bodies. Since physical bodies limit the levels of intelligence and awareness of the physical world that the minds occupy, in order for our souls to progress during subsequent incarnations *we needed increasingly complex bodies to do that.* We now have brains with capacity suitable for more highly developed souls—if we choose to continue growing.

Evidence has been presented in séances and hypnotic regressions that indicate during the period between physical births, the soul reviews previous lives to learn what was done properly and what was done wrong. This concept of change while in the afterlife is consistent with the concept of Purgatory. (Of course, none of the mainline Christian denominations go on to the next step—reincarnation.) After the soul in the afterlife has learned its lessons, a new body is selected for rebirth and continued growth. The soul, with help from more advanced spiritual advisors, chooses a physical body that will be in the conditions and environment that will encourage further spiritual growth.

While in physical bodies, we are expected to learn how to use our spiritual energy, preparing us for our ultimate destiny. The goal is spiritual perfection. We are still in the process of creation. The destiny for our perfected souls is union with God, as we become Sons of God. When we are perfect, we will be able to do the things God's Sons have done and continue to do.

Love Your Neighbor

The Old Testament tells us how God dealt with Hebrews that had advanced spiritually farther than most others of their time. They were molded by revelations, persuasion, and persecution into the people that became an inspiration for the rest of the world. Christians like to think their religion is more advanced than its parent Judaism is, whether that is true or not. Although compared to most other religions of the world, most of what is called Christianity is not much more developed than the pagan religions replaced by it. Nevertheless, many Christians appear to be more spiritually advanced than members of many other religions are.

Love for everyone is the message that God, through Jesus Christ, has tried to give people for many centuries. When souls do become Sons of God, they will have to love one another so they can cooperate in the task of *mending the world.*

GOD

WHAT DOES GOD DO?

God is the name for the total being of creation. God is singular, monotheistic; but God is a celestial corporation, the assembly of all the advanced spiritual beings that contribute to the activities that proceed in the entire Universe. God works through and with what Christians call Father, Son, and Holy Spirit. Between them, they are *the* God of the entire Universe. This is not a polytheistic concept; the spirits that can function as individuals operate in perfect harmony. They have perfect rapport. They have a complete power of telepathy. What one knows, they all know. They can all function as one being; or only one or a few can operate independently of the others, even though they all are aware of what each partner does. Parts of this plural God created the Solar System and human beings; but the entire being of God was aware of what was happening in this obscure part of the entire creation. The spirit in the body of Jesus was a partner in our creation, done with a single, corporate thought and a word. "*And **God** said Let there be light and there was light.*" (Genesis 1:3)

The Bible and Creation: The story of creation in the Bible states that the word *God* includes multiple personalities, without stating how many were involved. However, the singular word, *God,* indicates that the multiple personalities worked together by using the plural pronoun when referring to creation of mankind. The idea of a corporation acting as an individual with individual members of the corporation acting on their own can be illustrated with a family business structure.

J. God and Son

Imagine a family business called, "J. God and Son." These two are partners and any one of them can speak for the company. However, the products being delivered from the company are sent by the marketing manager (the Holy Ghost), who has been with the company since it was established. He is involved with all decisions made by the owners and completely understands their philosophy. These three, representing the Father, Son, and Holy Ghost, each own one-third of the stock. Messengers (angels) were hired after the company was established and they perform various services to implement company policy, such as delivering messages and guarding and protecting customers of the organization. Another member of the business is a manager called Satan, who tests people to see if they really are customers of the business. Satan also destroys products that are being misused.

This example certainly does not explain the concept of a multiple monotheistic God; but it illustrates how a corporation operates. The God of the Universe functions something like that. I believe that the being that Christians call Jesus Christ is responsible for the creation and maintenance of our Solar System. This part of God actually appeared on Earth in the person of Jesus of Nazareth. This being also appeared as the individual that Hindus call Krishna. Both Mohammed and Zoroaster received revelations from spiritual beings: Mohammed from the angel *Gabriel*, and Zoroaster through *The Good Mind.* These beings functioned like what Christians call the Holy Ghost. The being called Satan will be discussed later.

The being that was in the body of Jesus of Nazareth in not named as an individual; but the being is called the Son of God. The Gospel of John (John 1:1-14) states the being that was with God at the beginning was *The Word.* This word was "*with God; but it also was God.*" This being was the light of the world. (John 8:12) The light came into the world in the body of Jesus of Nazareth. "*The Word was made flesh, and dwelt among us...*" Jesus was baptized at the age of about thirty, marking the beginning of his public ministry. At that time a voice from heaven said, "*This is my beloved Son, in whom I am well pleased.*" (Matthew 3:16, 17) (Notice that Jesus "*went up straightway out of the water,*" apparently in the same position his body was in when it was resurrected, and which was the position of Lazarus when he was brought back to life.) John the Baptist is recorded in John 1:32-34 testifying in one of his sermons that Jesus was the Son of God.

This Son came into the world to guide us onto the correct path to our destiny. (John 3:16-19) In the body of Jesus of Nazareth, the Son of God taught us that we must believe in the Son; this is necessary because we must be able to understand and accept what he told us. The message was that we *can* become Sons of God: "*as many as received him, to them gave he power to become the sons of God.*" However, we must also understand and obey his instructions given in the Gospels, especially the commands to love God and to love our neighbors. It is only by obedience that we prove our belief.

Matthew 3:16, 17 And Jesus, when he was baptized, **went up straightway out of the water**: and, lo, the heavens were opened unto him, and he saw the Spirit of God descending like a dove, and lighting upon him: And lo a voice from heaven, saying, **This is my beloved Son,** in whom I am well pleased.

John 1:32-34 And John bare record, saying, **I saw the Spirit descending from heaven like a dove, and it abode upon him**. And I knew him not: but he that sent me to baptize with water, the same said unto me, Upon whom thou shalt see the Spirit descending, and remaining on him, the same is he which baptizeth with the Holy Ghost. And **I saw, and bare record that this is the Son of God**.

John 1:1-14 In the **beginning** was the **Word**, and the Word was with God, and the **Word was God**. The same was in the beginning with God. All things were made by him; and without him was not any thing made that was made. In him was life; and the life was the light of men... That was the true Light, which lighteth every man that cometh into the world. He was in the world, and the world was made by him, and the world knew him not...But **as many as received him, to them gave he power to become the sons of God,** even to them that believe on his name: Which were born, not of blood, nor of the will of the flesh, nor of the will of man, but of God. **And the Word was made flesh, and dwelt among us...**

John 8:12 Then spake Jesus **again unto them, saying, I am the light of the world**: he that followeth me shall not walk in darkness, but shall have the light of life.

John 3:16-19 For **God so loved the world, that he gave his only begotten Son,** that whosoever believeth in him should not perish, but have everlasting life. For God sent not his Son into the world to condemn the world; but that the world through him might be saved. He that believeth on him is not condemned: but he that believeth not is condemned already, because he hath not believed in the name of the only begotten Son of God. And this is the condemnation, that light is come into the world, and men loved darkness rather than light, because their deeds were evil.

Sons of God

At the beginning of the Universe, there were beings, which seem to be subservient to a higher authority, perhaps the one called "Father" by Jesus. (As mentioned earlier, the term Father might refer to the corporate being of God, instead of to just one member of the Godhead.) The description of the creation process in the Bible does not identify these beings by name; it merely uses the plural pronoun for God, as discussed earlier. The names we use for them are inferred from statements in various parts of the Bible; and they are called, "Son of God," "Holy Spirit," and/or "Angel." Some places in the Bible these terms indicate different beings or different functions, other places it is difficult to determine if there is a difference between them. Representatives of the groups of subservient beings have appeared and continue to appear on Earth to perform actions or deliver messages from God. The only begotten Son of God was born on Earth as Jesus of Nazareth. He appeared on Earth to deliver a message from God, the Father. Jesus prayed to the Father and said that he obeyed the Father.

John 14:31 But that the world may know that **I love the Father; and as the Father gave me commandment, even so I do**. Arise, let us go hence.

Jesus also said the Father was greater than he was. In the following passage from the gospel of John, Jesus introduces another being, the *Holy Ghost*, which has a function not previously discussed in the Bible. In some ways, the Holy Ghost acts like a superior angel, but its description indicates an entirely new and separate being.

> **John 14:25 28** These things have I spoken unto you, being yet present with you. But the Comforter, which is **the Holy Ghost**, whom the Father will send in my name, **he shall teach you all things**, and bring all things to your remembrance, whatsoever I have said unto you. Peace I leave with you, my peace I give unto you: not as the world giveth, give I unto you. Let not your heart be troubled, neither let it be afraid. Ye have heard how I said unto you, I go away, and come again unto you. If ye loved me, ye would rejoice, because I said, I go unto the Father: for **my Father is greater than I**.

Adding to the confusion, the terms, "Angel of God" and "Angel of the Lord" used in several places seem at some times to refer to Yahweh himself, other times to a son or the Son of God, other times, to a high-ranking angel. Many Christian Bible scholars teach that the term "sons of God" in the following passages actually means angels. However, Satan seems to have more authority than usually ascribed to ordinary angels—he had independent thoughts, as if he were a Son of God.

> **Job 1:6** Now there was a day when the **sons of God** came to present themselves before the LORD, and **Satan came also among them**.

> **Job 38:1-7** Then the LORD answered Job out of the whirlwind, and said, Who is this that darkeneth counsel by words without knowledge?...Where wast thou when I laid the foundations of the earth? declare, if thou hast understanding...When the **morning stars** sang together, and **all the sons of God shouted for joy**?

The book of Job is consistent in its reference to the plural "sons," confirming use of the plural "us" in the book of Genesis describing the creation of mankind.

The following passage is even more confusing. Here, we are told that the "sons of God" *"took them wives of all which they chose."* However, Matthew 27:30 states that angels *"marry not nor are they given in marriage."*

> **Genesis 6:1-2** And it came to pass when men began to multiply on the face of the earth, and daughters were born unto them, That **the sons of God saw the daughters of men that they were fair; and they took them wives** of all which they chose.

In the following passage from Daniel, the term, "Son of God" evidently refers to a being that was of a higher order than angels. Christian scholars believe this refers to the pre incarnate Christ. However, the being described below is performing a task that is usually assigned to angels—protecting humans from harm.

> **Daniel 3:20-25** And he commanded the most mighty men that were in his army to bind Shadrach, Meshach, and Abednego, and to cast them into the burning fiery furnace. Then these men were bound in their coats, their hosen, and their hats, and their other garments, and were cast into the midst of the burning fiery furnace...Then Nebuchadnezzar the king was astonied, and rose up in haste, and spake, and said unto his counsellors, Did not we cast three men bound into the midst of the fire? They answered and said unto the king, True, O king. He answered and said, Lo, I see four men loose, walking in the midst of the fire, and they have no hurt; **and the form of the fourth is like the Son of God.**

The Holy Three

Many religions of the past and present, in addition to Christianity, have some form of multiple Gods, usually in groups of three that share some characteristics with those the Bible ascribes to God.

Ancient Egyptian beliefs: The Supreme Being was called Ra, represented by the Sun. He had two other forms, Aton and Khepri that helped him create the world. Various parts of Egypt had different names for the three aspects of God; but they all believed in the same kind of creative triad.

Norse mythology: The entire Universe was created from the combination of intense heat and frozen water. As water melted, a complicated process resulted in the birth of three sons, Odin, Vili, and Ve, who between them created the world and humans.

Greek mythology has several stories about creation. One of them came from the Orphic cult, which was a major religion of ancient Greece. In this story, Chronos, the personification of time, created Phanes, who had many forms, including Eros, the god of love, and his daughter. These three created earth and all living beings.

Hinduism teaches there are many Gods; but three of them are the most important. Brahma is the greatest and unborn. Vishnu is part of the Godhead and is responsible for helping humans to find truth. He has appeared on earth in several different forms. The most recent appearance was in the form of a human called Krishna. He will return one more time. The third member is called Shiva who is the destroyer of mankind when their evil becomes too great to be allowed to continue.

Shinto teaches belief in "Kami," meaning spiritual beings. Before the beginning, there were three Kami, including "The Lord of central heaven" and two other beings that participated in the creation of all things. Eventually many other Kami were formed, culminating in the creation of humans. Amaterasu, the Sun Goddess, is the most important Kami. Her brother left the celestial place and traveled around Earth.

Taoism is founded on the concept of the Tao, which is the power surrounding and within all things, living and non-living. It is the cause of creation, the force or spirit of all living things. The Tao is not exactly a God in the sense of Western religions; it is more like the "Absolute" of Buddhism. The Tao is the source of the One. The One is the source of the Two. The two gave birth to the Three. The Three gave birth to all living beings.

Judaism claims to be strictly monotheistic. However, in Genesis 1:26 is a statement made by God: *"let us make man in our image,"* using plural pronouns. The Old Testament includes many references to "sons of God" and angels. It is not clear anywhere what these terms mean; but it seems there are either three groups or three individuals that share the kingdom of Heaven.

Zoroastrianism teaches a message of one God, Ahura Mazda (the Wise Lord), rejecting worship of all other gods. However, Ahura Mazda had twin sons who participated in creation of the world: one is the Holy Spirit who chose the good attributes: truth, justice, and life; the other son is the Evil Spirit who chose the forces of falsehood, injustice, and death.

Islam teaches there is only one God, Allah, who must be worshipped. However, Mohammed received his revelations from a spiritual being he called the angel Gabriel. In addition, there is an evil being called Shaitan who rules over demons.

Polynesian beliefs: Before being affected by contact with Western religions in the eighteenth century Polynesians believed in an invisible uncreated Supreme Being that was the ruler of the Universe. This Being created all things, and had a son and a third member of the Godhead that was their messenger on Earth. It is possible that some early Christian missionaries had reached the islands and brought this concept with them. Nevertheless, this concept is part of most primitive Polynesian cultures.

SATAN

Christians believe Satan, although identified as a fallen angel, is as much an individual personality as members of the Holy Trinity. Satan might have been an archangel before he acquired his present characteristic. In his present state, he is the chief or ruler of other fallen angels. Satan, or some being pretty much like him or her, has been demonized, literally, in most societies for thousands of years. Islam and Hinduism, as well as Jews and Christians, recognize the idea of a powerful spiritual being that causes harm to humans and human societies. Satan, in religions that recognize existence of this being is usually considered to either be a male or have a masculine personality, although in Hinduism and some legends, the local concept of Satan includes a female relative or spouse that is just as threatening as he is.

Satan has been considered the adversary of mankind. The New Testament from Matthew to Revelation uses the word devil more than 50 times when referring to Satan. 1Peter 5:8 identifies the adversary as the devil.

> **1Peter 5:8** Be sober, be vigilant; because **your adversary the devil**, as a roaring lion, walketh about, seeking whom he may devour.

Revelation 20:2 combines "*the dragon, that old serpent* and *devil*" when referring to Satan. Revelation 12:7-9 describes a war in heaven and predicts that Satan and his angels will be *cast out into the earth, and his angels* (will be) *cast out with him*

> **Revelation 20:2** And he laid hold on **the dragon, that old serpent**, which is the **Devil,** and **Satan**, and bound him a thousand years,

> **Revelation 12:7-9** And there was war in heaven: Michael and his angels fought against the dragon; and the dragon fought and his angels, And prevailed not; neither was their place found any more in heaven. And the great dragon was cast out, that old serpent, called the Devil, and Satan, which deceiveth the whole world: he was **cast out into the earth, and his angels were cast out with him**.

Satan will be treated harshly at the end times. This is prophesized in at least three different books of the Bible.

John 12:31 Now is the judgment of this world: now shall **the prince of this world** be cast out.

Daniel 7:11 I beheld then because of the voice of the great words which the horn spake: I beheld even till **the beast was slain**, and his body destroyed, and given to the burning flame.

Revelation 19:19, 20 And I saw **the beast, and the kings of the earth**, and their armies, gathered together to make war against him that sat on the horse, and against his army. And **the beast was taken**, and with him the false prophet that wrought miracles before him, with which he deceived them that had received the mark of the beast, and them that worshipped his image. These both were cast alive into a lake of fire burning with brimstone.

Beelzebub is another name used for Satan in the New Testament. (Matthew 10:25; 12:24, 27; Mark 3:22) Matthew 12:24 identifies Beelzebub as the "*prince of the devils.*" The word probably is a distortion of the word Baalzebub, the god of Ekron, which was a Philistine city. (2Kings 1:2, 3, 6, 16)

Matthew 10:25 It is enough for the disciple that he be as his master, and the servant as his lord. If **they have called the master of the house Beelzebub**, how much more shall they call them of his household?

Matthew 12:24 But when the Pharisees heard it, they said, This fellow doth not cast out devils, but by **Beelzebub the prince of the devils**.

Satan and Others like Him in Other Religions

I am exploring the concept of Satan because I think descriptions of this being give us some clues to the spiritual evolution and destiny of humans. Our evolution has been a battle against Satan's temptations, with some humans victorious, and many succumbing to his temptations. One of our objectives while in physical bodies on Earth is to overcome temptations and to regret and repent for the times we failed the tests. Many, probably most, religions have some spiritual being that has characteristics similar to those Christians believe are attributed to Satan. By examining the various concepts, I think it is possible to get a clearer understanding of the role the Satan character has in the spiritual evolution and advancement of human souls.

Zoroastrianism, Angra Mainyu: At the beginning of creation, the twin sons of *Ahura Mazda* started an eternal rivalry. *Spenta Mainyu* (Holy Spirit) chose the good attributes: truth, justice, and life. *Angra Mainyu* (Evil Spirit), the other son, chose the forces of falsehood, injustice, and death. The evil son became an opponent of Ahura Mazda. In this position he is called *Ahriman.* He rules a kingdom of the Lie, ruling over evil spirits. Human beings are participants in the celestial battle between these two spiritual beings. Eventually the evil one will be overcome, proving the superiority of truth, justice, and life.

There are many teachings from Zoroaster that are part of the Christian religions. Spenta Mainyu is titled Holy Spirit; but this character is more like the person of Jesus Christ. There is no question that Angra Mainyu is like Satan. In both Zoroastrianism and Christianity, humans choose to support either the evil one or the good one. Both religions teach about a final battle with Satan, who is in a celestial realm, in which humans on Earth participate.

Ancient Egyptian Religion: Osiris, Lord of the Underworld: In very ancient times, Osiris was the ruler of Egypt and the rest of the world. He was the great-grandson of Ra the sun god, who had returned to the Sun, leaving the position of supreme ruler of Earth to his son, Shu, leading to the birth of Osiris.

Osiris was so full of love and goodness, that he did not realize his brother, Seth, was extremely jealous of him. Eventually, the jealousy worked itself into action, and Seth killed his brother. After the murder, the body was dismembered and parts scattered around Egypt. Isis, the wife and sister of Osiris, found the parts and reassembled them. However, he had been in the world of the dead too long to return to the land of the living, so he became Lord of the Underworld.

He was a wonderful being, closely related to the chief god. He became the tester of the dead that had journeyed to his kingdom below the Earth. After reaching the presence of Osiris, those who were found as enemies of Ra and not worthy of joining him forever, were either burned or boiled to final extinction. If the soul was deemed worthy, the journey continued toward a narrow gorge that led to the god Ra. On this journey, many dangers, including many demons, threatened the soul. Egyptians studied The *Book of the Dead* to learn how to survive that treacherous journey.

Because of Osiris' role in maintaining fertility of the Earth and to the final success of souls, cults sprang up to worship him. These cults were in competition with other cults, taking power and wealth from priests of other forms of worship. In another parallel with modern religions, Satan has been and still is worshipped by many who find the main organized religions uncomfortable or unworthy of their time and support.

Many of the characteristics of Osiris resemble features of Satan and his position in the religions that evolved from that of Egypt. (The Hebrew religion borrowed many features of the Egyptian religions. The temple and the role of the priests were similar to those of Egypt. Evidently, Moses had been familiar with at least one of the cults of the land where he spent 40 years.) Both Osiris and Satan were once highly regarded, but were removed from power—Satan because of his jealousy of God, Osiris because of his brother's jealousy. Both were sent to the underworld. Osiris tests the souls of the dead. Satan tempts the souls of the living. One of the symbols associated with Osiris is the cobra snake. Satan has been compared to a serpent. However, Osiris does not cause difficulties for the living in the same manner that Satan is supposed to do.

On the other hand, Osiris is also like Jesus Christ. He was killed by his enemy and was resurrected. He was transported to the underworld, but still survives in a spiritual realm to greet newly dead humans. Many people, not only Christians, believe someone or something that has been identified as Jesus Christ by Christians interviews and tests people who have died physically.

Hinduism, Shiva: Hinduism recognizes many gods, one of which, Shiva, is a member of the Hindu triad. Shiva is the god of destruction *and* regeneration. Many of his characteristics are shared with Osiris and Satan. They all are gods of death. All three are involved with demons that share the underground. Shiva has a wife who has four characteristics—two which can be creative and two which are destructive. The destructive characteristic includes fearful demons. The most important function of Shiva is to destroy societies that have become too evil to be allowed to survive. It is not clear how the distinction is made when to destroy a society.

While there are some similarities between Shiva and Satan, nowhere in the Holy Bible is Satan specifically given responsibility to destroy societies. When Sodom was destroyed, angels were responsible. (Genesis 18:1-25) Of course, Satan is supposed to be a fallen angel; so he or other fallen angels might have done that. However, Satan has great power over humans and Earth, but only if humans allow him to do that. (The power and authority of Satan is discussed later in the following sections.)

Norse Mythology, Loki: Norse mythology includes a character named Loki. He had a monstrous daughter, named Hel, who is in charge of the underworld. Loki has many characteristics similar to Satan, Angra Mainyu, Osiris, and Shiva. Loki was at first the friend of Odin, the chief Norse god. However, he eventually revealed his true nature and became an enemy of the pantheon of gods. He was conquered and bound, but not destroyed because he is destined to participate in the final battle between good and evil.

Loki is more closely related to Shiva than to Satan; because of the relation to a female that has monstrous characteristics. The name Hel, as the daughter of Loki, evidently is where Europeans got the name for the underworld where some dead humans continue to exist, because they recognized the similarity when they were introduced to Christianity. (Hades, another word for the place where dead humans reside, is from the Greek language.)

Islam, Shaitan: Islam is based on five doctrines, one of which is Angels exist. The chief angel is Gabriel. There is also a fallen Angel named Shaitan with his followers, the djinn (demons). This being is quite similar to the Judeo-Christian Satan and is frequently called by that name by English-speaking Muslims. The resemblance is no coincidence; Mohammed had learned much about both Judaism and Christianity before he began preaching his unique message and before he began to hate and persecute Jews.

Non-theistic Religions, Other Evil Beings: For the Anti-Theists and some Atheists, God of the Judeo-Christian religion is the evil force in this world. Communists, and other Socialists believe capitalism and capitalists are evil, replacing Satan as the source of human suffering. Hitler and his brood of underlings were the evil force in Europe for almost 20 years. In the USA, Republicans represent unmitigated evil for some Democrats and vice-versa. These varying opinions may be simply a way for moderns to explain the existence of evil and suffering in the world, or they might even be true in some cases. In fact, belief in a spiritual evil being may also be man's attempt to understand nature or to transfer responsibility for evil from himself to some imaginary being or force.

The Fallen Angel

I believe there is some reason why Satan is mentioned so many times in the Bible; and why his counterparts are present in legends and documents of most religions. The Bible identifies Satan as the adversary of God and mankind, and he is lord of evil. He is also the Devil, supreme master of evil, lord of other fallen angels that joined him in rebellion against God. The fallen angels are supposed to be in Hell, a spiritual cavern inside the Earth.

> **2Peter 2:4** For if God spared not **the angels that sinned,** but **cast them down to hell,** and delivered them into chains of darkness, to be reserved unto judgment;

Satan was cast to the ground—a *fallen* angel. This could mean that he was sent to Earth from heaven either to punish him—or for the purpose of *tempting* mankind. The book of Job indicates that Satan is responsible for temptation and punishment. In the book of Job, *Satan was with the sons of God* during some kind of meeting. The implication is that Satan is, or was, one of the Sons. However, since Satan is said to be a fallen angel, what was he doing at a meeting where the "*sons of God came to present themselves before the Lord, and Satan came also among them*?" In order to explain the apparent anomaly, some theologians claim "sons of god" means "angels," not sons. The Lord carried on an amiable discussion with Satan, indicating that Satan was not an enemy of God. Nevertheless, most humans believe Satan is the enemy of human beings, because he tempts and punishes those who give in to his temptations.

> **Job 1:6-8** Now there was a day when the **sons of God came to present themselves before the Lord, and Satan came also among them**.
> And the Lord said unto Satan, Whence comest thou?
> Then Satan answered the Lord, and said, From going to and fro in the earth, and from walking up and down in it.
> And the Lord said unto Satan, Hast thou considered my servant Job, that there is none like him in the earth, a perfect and an upright man, one that feareth God, and escheweth evil?

(Why does God bring this up? Is he taunting Satan? Or is God ***telling*** *Satan what to do?)*

> **Job 1:9-12** Then Satan answered the Lord, and said, Doth Job fear God for nought? Hast not thou made an hedge about him, and about his house, and about all that he hath on every side? thou hast blessed the work of his hands, and his substance is increased in the land. But put forth thine hand now, and touch all that he hath, and he will curse thee to thy face.
> And the Lord said unto Satan, Behold, all that he hath is in thy power; only upon himself put not forth thine hand. So Satan went forth from the presence of the Lord.

(God gave Satan permission to tempt Job. Did God know that Job would remain true? Was God trying to demonstrate the weakness of Satan? Did God think that Job should learn some kind of lesson during the testing that he would suffer?)

Satan had been *"going to and fro in the earth, and from walking up and down in it."* If Satan was an outcast, it seems strange that he would have the freedom of doing all that walking around, and be allowed into a meeting with the Sons of God.

In the temptation of Jesus, (Luke 4:2-8) Satan claims to have power to give or take the kingdoms of the world. Some theologians think that this was an empty claim, since they think that Satan was lying. Maybe, but the Bible shows Satan is very powerful in the Earth. If Satan were lying or exaggerating, and if Satan were not capable of assigning power to Jesus, Jesus would have reminded him of that fact. However, Jesus told Satan that he would not accept the challenges he was tempted by.

> **Luke 4:2-8** Being forty days tempted of the devil. And in those days he did eat nothing: and when they were ended, he afterward hungered. And the devil said unto him, If thou be the Son of God, command this stone that it be made bread.
>
> And Jesus answered him, saying, It is written, That man shall not live by bread alone, but by every word of God.
>
> And the devil, taking him up into an high mountain, shewed unto him all the kingdoms of the world in a moment of time. And the devil said unto him, **All this power will I give thee, and the glory of them: for that is delivered unto me; and to whomsoever I will I give it. If thou therefore wilt worship me, all shall be thine.**
>
> And Jesus answered and said unto him, Get thee behind me, Satan: for it is written, Thou shalt worship the Lord thy God, and him only shalt thou serve.

Satan has been represented by a serpent, which lives in constant and complete contact with the Earth. Even after the war in heaven, (Revelation 12:9) Satan will be "*cast out into the earth.*" Many of us believe that Hell, the kingdom of Satan, is somewhere down below—inside the Earth. Therefore, it seems that Satan probably is part of the Earth.

> **Genesis 3:1-5** Now the serpent was more subtil than any beast of the field which the Lord God had made. And he said unto the woman, Yea, hath God said, Ye shall not eat of every tree of the garden? And the woman said unto the serpent, We may eat of the fruit of the trees of the garden: But of the fruit of the tree which is in the midst of the garden, God hath said, Ye shall not eat of it, neither shall ye touch it, lest ye die. And the serpent said unto the woman, Ye shall not surely die: For God doth know that in the day ye eat thereof, then your eyes shall be opened, and ye shall be as gods, knowing good and evil.

Prince of this World

The Bible states that Satan is the prince of this world, "*god of this world,*" and the "*prince of the powers of the air,*" confirming that Satan has power over "*this world.*"

2Corinthians 4:4 In whom the **god of this world** hath blinded the minds of them which believe not, lest the light of the glorious gospel of Christ, who is the image of God, should shine unto them.

Ephesians 2:1-3 And you hath he quickened, who were dead in trespasses and sins; Wherein in time past ye walked according to the course of this world, according to the **prince of the power of the air**, the spirit that now worketh in the children of disobedience: Among whom also we all had our conversation in times past in the lusts of our flesh, fulfilling the desires of the flesh and of the mind; and were by nature the children of wrath, even as others.

John 12:31 Now is the judgment of this world: now shall **the prince of this world** be cast out.

John 14:29,30. And now I have told you before it come to pass, that, when it is come to pass, ye might believe. Hereafter I will not talk much with you: for **the prince of this world** cometh, and hath nothing in me.

The title "prince" in the Bible, in reference to human authorities, refers to governors or officers of provinces. However, the book of Daniel calls the Messiah a prince. The book of Acts calls Jesus a prince, confirming the relation of Jesus and the Messiah, as well as identifying the authority of Jesus. Therefore, Jesus is the Prince of his Church, which is all those who believe in him.

Daniel 9:25-15 Know therefore and understand, that from the going forth of the commandment to restore and to build Jerusalem unto the **Messiah the Prince** shall be seven weeks, and threescore and two weeks: the street shall be built again, and the wall, even in troublous times. The God of Abraham, and of Isaac, and of Jacob, the God of our fathers, hath glorified his Son Jesus; whom ye delivered up, and denied him in the presence of Pilate, when he was determined to let him go. But ye denied the Holy One and the Just, and desired a murderer to be granted unto you; And killed the Prince of life, whom God hath raised from the dead; whereof we are witnesses.

Acts 5:30, 31 The God of our fathers raised up Jesus, whom ye slew and hanged on a tree. Him hath God exalted with his right hand to be **a Prince and a Saviour**, for to give repentance to Israel, and forgiveness of sins.

The question is, what does it mean to be a prince of this world? Since Satan is a fallen angel, he is on or in Earth. If he is a prince of the Earth, he must have power over what happens on Earth. Indeed, the previous discussion of his power indicates that Satan does have control of Earth and over what happens here. He is not only the tempter, but also it seems that he is the destroyer as well— like Shiva, at least in that function. Satan tempts people to sin and then he destroys them if they succumb to his temptations.

As mentioned earlier, many theologians think the following passage from Isaiah refers to a human, Earthly ruler. If that is true, it still shows the relationship of Lucifer and Hell. Perhaps Isaiah was showing the similarity of the ruler and Satan.

> **Isaiah 14:12-15** How art thou fallen from heaven, **O Lucifer, son of the morning!** how art thou cut down to the ground, which didst weaken the nations! For thou hast said in thine heart, I will ascend into heaven, I will exalt my throne above the stars of God: I will sit also upon the mount of the congregation, in the sides of the north: I will ascend above the heights of the clouds; I will be like the most High. Yet thou shalt be brought down to hell, to the sides of the pit.

Regardless of the true relationship of Satan with God, Jesus warned against worshipping Satan, the prince of this world. During the temptation of Jesus in the desert, he repeated the commandment from the Old Testament to rebuke Satan.

> **Luke 4:8** And Jesus answered and said unto him, Get thee behind me, Satan: for it is written, Thou shalt worship the Lord thy God, and him only shalt thou serve.

Satan takes knowledge of God from those that do not stay with God and follow the teaching of Jesus.

> **Mark 4:15** And these are they by the way side, where the word is sown; but when they have heard, Satan cometh immediately, and taketh away the word that was sown in their hearts.

Temptation

It appears that Satan's primary function on Earth is to tempt human beings. We are tempted for several reasons: to help us recognize out weaknesses so we can improve our spiritual welfare, to improve our strength to resist future temptations, and to identify the people that are dedicated to God. Satan tested even Jesus, who resisted the temptations and went on to begin evangelizing the world for God.

> **Matthew 4:1-11** Then was Jesus led up of the Spirit into the wilderness to be tempted of the devil. And when he had fasted forty days and forty nights, he was afterward an hungred. And when the tempter came to him, he said, If thou be the Son of God, command that these stones be made bread.
>
> But he answered and said, It is written, Man shall not live by bread alone, but by every word that proceedeth out of the mouth of God.
>
> Then the devil taketh him up into the holy city, and setteth him on a pinnacle of the temple, And saith unto him, If thou be the Son of God, cast thyself down: for it is written, He shall give his angels charge concerning thee: and in their hands they shall bear thee up, lest at any time thou dash thy foot against a stone.
>
> Jesus said unto him, It is written again, Thou shalt not tempt the Lord thy God.
>
> Again, the devil taketh him up into an exceeding high mountain, and sheweth him all the kingdoms of the world, and the glory of them; And saith unto him, All these things will I give thee, if thou wilt fall down and worship me.
>
> Then saith Jesus unto him, Get thee hence, Satan: for it is written, Thou shalt worship the Lord thy God, and him only shalt thou serve.
>
> Then the devil leaveth him, and, behold, angels came and ministered unto him.

Probably the most familiar case of Satan tempting humans is in the book of Job. (Discussed earlier.) Satan acted as adversary, tempter, and executioner, attempting to shake the faith Job had in God. Job passed the test, and he provides an example for all of us when we are tempted to disobey the commandments of God. Few of us ever are or will be tempted the way Job was tempted.

The Apostles were tempted in the Garden of Gethsemane; and

they failed the test. (Luke 22:31-40) However, eventually they recognized their failings; and they went on to become leaders of the early Christian Church. Judas had been tempted earlier and he failed the test and betrayed Jesus. Jesus would have forgiven him, if he had gone to the cross and confessed his failings; but he compounded his sins by killing himself.

> **Luke 22:31-40** And the Lord said, Simon, Simon, behold, **Satan hath desired to have you, that he may sift you as wheat**: But I have prayed for thee, that thy faith fail not: and when thou art converted, strengthen thy brethren.
>
> And he said unto him, Lord, I am ready to go with thee, both into prison, and to death.
>
> And he said, I tell thee, Peter, the cock shall not crow this day, before that thou shalt thrice deny that thou knowest me. And he said unto them, When I sent you without purse, and scrip, and shoes, lacked ye any thing? And they said, Nothing. Then said he unto them, But now, he that hath a purse, let him take it, and likewise his scrip: and he that hath no sword, let him sell his garment, and buy one. For I say unto you, that this that is written must yet be accomplished in me, And he was reckoned among the transgressors: for the things concerning me have an end.
>
> And they said, Lord, behold, here are two swords.
>
> And he said unto them, It is enough. And he came out, and went, as he was wont, to the mount of Olives; and his disciples also followed him. And when he was at the place, he said unto them, **Pray that ye enter not into temptation.**

> **Luke 22:3, 4 Then entered Satan into Judas surnamed Iscariot**, being of the number of the twelve. And he went his way, and communed with the chief priests and captains, how he might betray him unto them.

Peter reminds believers in Jesus Christ that even though they are tempted, they must not give in to it—if they want to receive "*the salvation of your souls.*"

> **1Peter 1:3-9** Blessed be the God and Father of our Lord Jesus Christ, which according to his abundant mercy hath begotten us again unto a lively hope by the resurrection of Jesus Christ from the dead, To an inheritance incorruptible, and undefiled, and that fadeth not away, reserved in heaven for you, Who are kept by the power of God through faith unto salvation ready to be revealed in the last time. Wherein ye

greatly rejoice, though now for a season, if need be, **ye are in heaviness through manifold temptations:** That the trial of your faith, being much more precious than of gold that perisheth, though it be tried with fire, might be found unto praise and honour and glory at the appearing of Jesus Christ: Whom having not seen, ye love; in whom, though now ye see him not, yet believing, ye rejoice with joy unspeakable and full of glory: Receiving the end of your faith, even **the salvation of your souls**.

Deuteronomy 8:1-3 reminded the Hebrews that those who survived the temptations and trials of the desert were humbled and began to trust the Lord. (It seems that humans cannot admit they need and trust someone, including God, unless they recognize their own weaknesses—be humble.)

Deuteronomy 8:1-3 All the commandments which I command thee this day shall ye observe to do, that ye may live, and multiply, and go in and possess the land which the Lord sware unto your fathers. And thou shalt remember all the way which the Lord thy God led thee these forty years in the wilderness, to humble thee, and **to prove thee, to know what was in thine heart, whether thou wouldest keep his commandments, or no.** And he humbled thee, and suffered thee to hunger, and fed thee with manna, which thou knewest not, neither did thy fathers know; that he might make thee know that man doth not live by bread only, but by every word that proceedeth out of the mouth of the Lord doth man live.

Genesis 22:1 states that God tempted Abraham; but Abraham followed God's directions and passed the test. 2Samuel 24:1 states that *God* tempted David; and when David succumbed to temptation, the entire nation of Israel suffered with a pestilence. (However, 1Chronicles 21:1 contradicts the verse in 2Samuel and states, "S*atan stood up against Israel, and provoked David to number Israel.*") 2Chronicles 32:30, 31 states that God tried Hezekiah in order to "*know all that was in his heart.*" However, James 1:12-15 states that God does not tempt humans. James declares that humans are "*drawn by their own lust.*"

Genesis 22:1 And it came to pass after these things, that God did tempt Abraham, and said unto him, Abraham: and he said, Behold, here I am.

2Samuel 24:1 And again the anger of the Lord was kindled against Israel, and he moved David against them to say, Go, number Israel and Judah.

1Chronicles 21:1 And Satan stood up against Israel, and provoked David to number Israel.

2Chronicles 32:30, 31 This same Hezekiah also stopped the upper watercourse of Gihon, and brought it straight down to the west side of the city of David. And Hezekiah prospered in all his works. Howbeit in the business of the ambassadors of the princes of Babylon, who sent unto him to inquire of the wonder that was done in the land, **God left him, to try him, that he might know all that was in his heart.**

James goes on to tell us that by giving in to temptation, our lust "*bringeth forth death.*"

James 1:12-15 Blessed is the man that endureth temptation: for when he is tried, he shall receive the crown of life, which the Lord hath promised to them that love him. Let no man say when he is tempted, I am tempted of God: for God cannot be tempted with evil, neither tempteth he any man: But every man is tempted, when he is **drawn away of his own lust**, and enticed. Then when lust hath conceived, it bringeth forth sin: and sin, when it is finished, **bringeth forth death**.

These seemingly conflicting passages and the material from the book of Job indicate that *Satan performs a task for God*—testing humans in order to prove that we really believe in and trust God. Therefore, either Satan and his position in the hierarchy of Heaven are part of God's plan to create pr develop Sons of God, or God is taking advantage of the fallen angel to eliminate humans that are not worthy of becoming Sons of God.

SUMMARY AND CONCLUSIONS

The Universe was not started by some freak accident. There is no way the laws of chance can be used to defend the concept of a cosmological accident. Creation of the Universe might have been a natural occurrence; but what and where is the source of nature? As far as we know, what we call "nature" was created at the beginning of the Universe. Every thing that modern science has taught us is that events have causes and consequences. The fact that the Universe does exist *proves* there was a cause. It is sensible to look for the simplest explanation for any dilemma. The simplest explanation for creation is that God did it. It is necessary to have faith that God exists. It is also necessary to have faith that accidents, if that is what you choose to believe in instead of in God, can cause all the sequential events that led to the creation and development of human beings.

God does exist. God created the Universe. God created all the living beings that inhabit the world. The world is real. Everything that happens here has a cause and some kind of consequence.

The Purpose of Creation

Most of us accept that the Universe exists and is as real as our own existence; but we do not know *why* it exists. Any explanation for existence of the Universe has to be accepted on faith, whether in "intelligent design" or in some secular theory. As I mentioned earlier, it is impossible to prove a negative; therefore, Atheists who claim they *know* there is no God or a source of intelligent design are not being very scientific nor are they very logical.

I believe in God, along with millions of other human beings. I think it is extremely important for us to believe in God. If we believe in God, we will accept that the Universe has some purpose. If the Universe has some purpose, living beings must also have some purpose. Therefore, we have responsibility to learn what that purpose is and what we should do about it.

It seems, according to the physical evidence, and to moral and religious history, that the most important product of creation is Human Beings, and/or some beings like us. We are the purpose for the entire Universe. Therefore, every individual human being has a purpose. In that case, we must allow each individual freedom to achieve that purpose in his or her way.

According to most religious teaching, Humans must try to draw close to the Creator. We must be perfect before we can be in the presence of God. Most of us want to achieve spiritual perfection, even if most of us do not use that term. Since it seems that few, if any, of us are perfect, there might be a method for us to reach that level. I think that we can reach perfection through evolution of physical bodies and evolution and reincarnation of spiritual souls.

The Purpose of Evolution

The evolution of Humans to our present mental and spiritual status started when God breathed spiritual life into one of the animals on Earth and made it into a *living* soul. This started a new form of evolution. Spiritual evolution continues the process of creation. From this point on, Humans as individuals or as societies can participate in the process, stagnate, or become extinct. Evolution of the physical bodies to contain brains that are more complex is the process by which creation has moved humans to higher and higher spiritual levels.

Animals evolved from small, microscopic organisms to the huge dinosaurs and finally to the animals that became Human Beings. On this purely physical growth of evolving, the forms that did not progress and improve stagnated or ceased to have descendants: they became extinct. Examples of stagnation are found in Australia with the marsupials. Societies that stagnated are present in many parts of the world. Some of the stagnant societies resist any attempt to bring them into a more evolved spiritual level. Examples of extinct animal forms are found in the fossils buried below the surface of the Earth, where the remnants of extinct societies are also found.

Forces of instincts and environment determine the desires and activities of animals without addition of spiritual life. The less developed animals are, the less freedom they have to resist these forces. Animals, including too many Humans with animal natures, still live by these rules. The objective of physical evolution has been to develop an animal with the capacity to continue evolution in a different way. Once animals were given living souls, they were given a new nature: a spiritual nature. Desires and activities are now, or should be, directed toward continued spiritual evolution of all human beings. If, and when, advanced souls require bodies that are more suitable for continued advancement, God will develop (evolve) them.

The state of humans in the future depends on what individuals and groups do to and about one another right now. The future of individual souls depends on the choices they make in their lives. Humans have freedom to choose whether to continue spiritual evolution or not; and we must resist efforts of demagogues to keep us bound to their forms of religion. We must allow souls to develop from where they are. The future of mankind depends on freedom for as many people as possible.

LOVE, KNOWLEDGE, AND VIRTUE

Many people have reported experiences of clinically dying but returning to physical life after which they claimed they had seen a brilliant light that some Christians identified as Jesus Christ. Non-Christians also reported seeing the bright light; but they did not usually assign a name to it. Most of those who returned from apparent death had the feeling that they should live different kinds of lives: they believed that they should be more loving toward others and they should continue searching for knowledge of all kinds. These two activities are recurring themes in the Bible. It appears that we must continue growing emotionally and intellectually and be loving and searching for true knowledge.

The Importance of Love

In addition to the "Ten Commandments," God gave two other commandments to the Hebrews of Moses' time; and Jesus repeated them to his disciples. The first is to love God. The second is to love your neighbor. These commands are repeated in some form in many places throughout the Bible. Mark 12:30 includes all the various statements in one verse: "*And thou shalt love the Lord thy God* ***with all thy heart****, and with* ***all thy soul****, and with* ***all thy mind,*** *and with* ***all thy strength***."

All my heart means that we are emotionally attached to God. We know He loves us. We need Him and trust Him. We want to be close to Him and His people and hope to join Him in His Kingdom of Heaven. We want to do His will and enjoy doing it. We enjoy being with others who believe in God. We worship God.

All my soul means that we are spiritually in harmony with God, and recognize that communication with Him and obeying His commands are important. We pray frequently and accept his spiritual guidance and control over our lives. And by saying the word "soul," we remind ourselves that we have souls and can continue to be with God forever, if we act as if we want to.

All my mind means that we are intellectually convinced that God is the Creator, that He is worthy of our love. We strive to learn God's will for us. We read and study the Bible and other religious material about God. Mark 12:33 in the King James Version repeats the words of this group but instead of saying "mind" it says "understanding," which means that we love God as much as we can with our limited ability to understand the message of God.

All my strength can also be stated "all my body." It means that we add our hands, feet, muscle, finances, and other physical resources to our devotion. Works are just as important as faith in our lives. We pray frequently. We praise God frequently. As we learn God's will for us, we do it willingly and happily because we love Him. In Deuteronomy 10:12 this is stated clearly in the words: *"walk in all his ways...and serve the LORD thy God...."*

Knowledge

The first mention in the Bible about knowledge was the "*knowledge of good and evil.*" God warned Adam that if he gained that knowledge, he would die.

> **Genesis 2:9** And out of the ground made the Lord God to grow every tree that is pleasant to the sight, and good for food; the tree of life also in the midst of the garden, and the tree of **knowledge of good and evil**.

I find it difficult to believe that God would punish Adam and Eve for doing something that He knew they would do. I agree with some students of the Bible who think that the admonition not to eat of the tree of knowledge was a test of mankind's ability to survive on his own. In that case, being expelled from the Garden of Eden was not his punishment, but the graduation of Adam on his next step toward his destiny. I think Adam and Eve proved they were not robots made merely to please God. When they displayed independent thinking, they proved they could grow to become suitable members of the being of God. This growth has taken and will continue to take a long, long time by our standards. Eventually, the souls of humans that survive will understand the significance of obeying the rules of God and they will obey willingly because they know they can obey and continue to retain their own personalities. Living on Earth in physical bodies gives us experiences and knowledge that help us to grow and learn how to be Sons of God. Our souls will reincarnate and inhabit increasingly complex bodies as we draw closer to our destiny. The steps toward our destiny are similar to the public school system in most countries: except there are many more "grades" to "pass" before "graduating" to the final destination.

Acquiring knowledge is valuable only if it is truth—real truth, not a personal definition that is satisfying but leaves a person in more ignorance than if he or she had not attempted to acquire knowledge. Praying regularly with a mind receptive to the leadings of the Holy Spirit can help us to be *"rightly dividing the word of truth."*

> **2Timothy 2:15** Study to shew thyself approved unto God, a workman that needeth not to be ashamed, **rightly dividing the word of truth.**

Virtue

The following Bible passages mention the word "virtue." One definition of the word means *potency or a capacity to act.* This is energy. When he healed people, Jesus lost or consumed virtue (energy). He could feel this loss.

> **Luke 6:19** And the whole multitude sought to touch him: for **there went virtue out of him,** and healed them all.

> **Luke 8:43-46** And a woman having an issue of blood twelve years, which had spent all her living upon physicians, neither could be healed of any, Came behind him, and touched the border of his garment: and immediately her issue of blood stanched. And Jesus said, Who touched me? When all denied, Peter and they that were with him said, Master, the multitude throng thee and press thee, and sayest thou, Who touched me? And Jesus said, **Somebody hath touched me: for I perceive that virtue is gone out of me.**

Paranormal experiences and experiences with miracles, angels appearing out of nowhere, ghosts, and other spiritual inter-actions in human affairs raise doubt about the reality of our world. If miracles happen, and if angels do appear and provide some kind of material assistance, where do they come from and where do the objects they seem to use come from. How can they move solid objects if they are not physical beings? How do ghosts or spirits communicate with humans? Why and how does prayer work? (See discussion about angels and prayer in *DESTINY, Volume Three.*)

The answer to all these questions and the only one consistent with a real world is that humans provide the energy for spiritual beings to appear and act; and they can affect material objects by their thoughts. Many people who reported encounters with ghosts or spirits say they noticed a drop in temperature when they sensed or saw the ghosts. This probably represented a transfer of the heat energy from the living human being to the spirit, which caused the person to feel colder. Humans can either resist or allow the transfer of energy, which is why some people in a group report sensing ghostly presences while others in the same group cannot sense the presence of the ghost.

If the above explanation is the correct answer, that explains why group prayers seem to be more successful than prayers by solitary individuals—there is more total human energy. Jesus mentioned the importance of group prayer in Matthew 18:19, 20.

> **Mt 18:19, 20** Again I say unto you, That if two of you shall agree on Earth as touching any thing that they shall ask, it shall be done for them of my Father which is in heaven. For **where two or three are gathered together in my name, there am I in the midst of them**.

The following passages seem to be defining virtue as a form of superior goodness. That is a good definition, but the way the word is used, it could also mean accumulation of spiritual energy. In both references, virtue is related to knowledge. 'Knowledge is power," was the motto of the high school I attended. That may be true; but power is composed of energy *and* action. In electrical formulas, power (watts) is the product of voltage (energy) and current (flow of energy). Without energy, actions cannot be initiated. Without action, energy is merely stored capability. In Christianity, virtue is the energy; love is the action. The power of Christianity is the product of virtue and love.

> **2Peter 1:5-9** And beside this, giving all diligence, **add to your faith virtue; and to virtue knowledge**; And to knowledge temperance; and to temperance patience; and to patience godliness; And to godliness brotherly kindness; and to brotherly kindness charity (*love*). For if these things be in you, and abound, they make you that ye shall neither be barren nor unfruitful in the knowledge of our Lord Jesus Christ. But he that lacketh these things is blind, and cannot see afar off, and hath forgotten that he was purged from his old sins.

> **2Peter 1:2-4** Grace and peace be multiplied unto you through the knowledge of God, and of Jesus our Lord, According as his divine power hath given unto us all things that pertain unto life and godliness, through the knowledge of **him that hath called us to glory and virtue**: Whereby are given unto us exceeding great and precious promises: that by these ye might be partakers of the divine nature, having escaped the corruption that is in the world through lust.

Human beings can accumulate spiritual energy during life on Earth. Every good deed, every good thought, all loving thoughts, and all knowledge that is acquired will add to the total energy. When a soul has accumulated enough spiritual energy, it will advance to a higher spiritual level, and, if the growth continues, the soul eventually will not be reincarnated on Earth again. The ultimate goal is to become permanently a spiritual being, where advancement to higher spiritual existence can take place.

Demagogues and groups that encourage hate, anger, and ignorance cause a reduction in the level of accumulated energy, reversing the growth of their followers. Anger hurts. Laughing at misfortunes of other people hurts. Evil acts and causing harm use up energy, and do not rebuild it. It is wasted; similar to losing electrical energy in a short circuit. If energy is depleted too much, the soul can regress to lower physical bodies. If the decline continues, the soul dies.

Religions should be helping us to advance our levels of energy. Those that teach love and knowledge do. Love helps. Learning helps. Obeying the Ten Commandments and the commands to love God and our neighbors can prevent loss of energy; and a sincere adherence to all the commandments from God will increase our spiritual energy.

Discipline

Using virtue is similar to exercising a physical body. Spiritual exercise builds spiritual "muscle" in the same way physical exercise builds up physical muscle: but it does take effort and some amount of pain, just as in physical exercise: "No pain, no gain." Spiritual exercise, such as praying for other people, comforting other people, and other good deeds use up spiritual energy; but it is restored by prayer, sacrifice, and resting the mind and body.

The soul is the controlling factor in human lives. It has the authority to control all other parts of the body and mind. The body is simply animal, and it has all the needs and desires of other mammals. The soul is supposed to learn how to control the animal nature. By learning how to control and protect a biological body, a soul graduates to helping other souls to grow spiritually.

The Old Nature

The animal nature remains part of our physical bodies. We need physical bodies to house our souls while we grow spiritually; but we must continually strive to overcome the animal nature. (One of the advantages of believing that human bodies have evolved is that we recognize the need to overcome our inherited animal nature.) Our continued spiritual growth depends on how we use our free will. We can choose to grow spiritually; or we can submit to the animal nature that we must overcome if we choose to grow. Overcoming temptations of our animal past requires discipline. Discipline is mentioned in the Bible many places. The following verses from Job are typical of God's attitude toward discipline.

> **Job 36:10-14** He openeth also their ear to discipline, and commandeth that they return from iniquity. If they obey and serve him, they shall spend their days in prosperity, and their years in pleasures. But if they obey not, they shall perish by the sword, and they shall die without knowledge...They die in youth, and their life is among the unclean.

Dogs and some other animals can be trained by humans to overcome some animal characteristics; but many natural traits will remain. These animals are trained by adapting the nature of the animal to desires of humans. They do what they are taught to do because of desires for rewards and love or fear of their owners. They will return to wild behavior once these stimulants are gone; and descendants cannot profit from their parents' experiences.

Humans can be trained by adapting our nature to the desires of God. Humans must understand why certain things are done and when the actions are no longer required. For example, a dog will learn not to touch a hot stove once it has been burned. However, that dog will not touch *any* stove, hot or cold, after learning that lesson. Humans learn not to touch a hot stove, but we know how to distinguish between hot and cold stoves. We should learn from our own experiences and those of others that failure to adhere to the will of God is similar to not understanding that some stoves are hot and can cause severe pain when touched.

Humans can learn from many sources besides personal experiences: our parents, our extended families, from people that live far away from us, or those who have died centuries before us. This is possible because of our communication skills. We can talk. We can write and others can read our products. It can be stored for use by subsequent generations. Information becomes tradition if it is valuable and contributes to spiritual growth, material growth, and survival of a society. Traditions that do not contribute to growth and survival are destructive instead of beneficial.

"Tradition" is the tool for learning from the past. The stage play and movie, *Fiddler on the Roof* illustrates this. In the story, traditions are given credit for survival of the Jewish culture. The story dramatizes the danger of losing traditions. Once traditions are lost or ignored, a culture loses its value to descendants and to the rest of the world. Prior to the Christian era, Jews tried to isolate themselves from the rest of the world and stagnated. Jesus attempted to bring them into harmony with the will of God by preaching love for all mankind and was executed for his efforts. Christians accepted many of the Jewish traditions and built on them to establish a new society, which has contributed to advancement of the entire world.

The Decaying Universe

There are several theories about the nature of the Universe. One of them suggests that the material within the Universe will eventually collapse into a very small amount of space. On the other hand, the material within it might continue spreading farther and farther apart as it appears to be doing now. There are other theories about the end of the Universe; but either complete collapse or continued dispersal is the most likely to occur. What happens to the Universe depends on the total amount of material within the Universe. Gravitation of the entire Universe affects all the material, with the center of gravity at or near the place where the Big Bang took place. If the amount of material is great enough, the combined pull of gravity at the center of the Universe will eventually retract to that area, just as a ball that is thrown in the air by a human eventually falls back to the ground, regardless of how fast or high the human threw it. If the amount of material is not sufficient to do that, the Universe will continue to expand.

The big bang was probably an instantaneous "explosion;" and the energy used to create the Universe probably was limited to that occurrence. Much of it has been expended in creation of space objects, which are slowly being converted into energy, which probably is not capable of creating a gravitational pull. Therefore, there may not be sufficient material to create a gravity field strong enough to pull the Universe back together.

If new energy is not provided from somewhere or somehow, the amount of energy still in the Universe will gradually be spread evenly throughout space, until all matter is at the level of absolute zero, perhaps a degree or two higher, the final temperature depending on what is "outside" of the Universe.

I believe there is a plan or purpose for creation of the Universe and mankind and for our future existence, I believe the plan involves a return of all matter to the center of the Universe, when a new, or maybe more, universes will be created. The only way that we know of to do that is to have enough material in the Universe. That means that there must be a new source of energy and convert it to material. Either God or the Sons of God could provide that energy. We, all the living physical beings on Earth, must continue growing spiritually until we become part of that group and we—in harmony with all the other spiritual beings that have reached that level are able to combine our efforts into generating enough spiritual energy.

Being able to perform the wonders that are required to reconstruct the objects of our Universe means each soul must be willing and able to cooperate with other souls that are involved in the same act of creation and development of these celestial worlds. Learning how to cooperate in this life means following the command from both the Old and the New Testaments: *love your neighbor as you love yourself.*

Our Destiny

The destiny of human beings is to join God and become Sons of God. Some of us will reach that destination; and some will fall alongside the road to whatever, wherever, and however that might be. Those that reach the level of being with God will retain the same personalities that they have now, except their personalities will be much more knowledgeable and much more loving than they are now. They will change so much that the level where most of us are now will be a distant memory. They will be to their present selves with a difference much greater than the difference between a mature adult and a baby at the time of conception. The process that has taken our personalities to our present level of development will continue; we will continue to reincarnate and the bodies that we inhabit will evolve to accommodate the advanced souls that continue to live on Earth.

The resurrected body of Jesus Christ was recognizable as the man who had died on the cross. Jesus ate regular food and seemed able to use it properly. Yet, he was able to pass through walls; and in the end, he was taken up to a cloud and disappeared. Philippians promises that our bodies, too, will change into something similar to the body of Jesus. His body demonstrated that we will be recognized as the same personalities we are now; however, we will have different kinds of bodies that have different abilities.

John 20:19, 20 Then the same day at evening, being the first day of the week, when the doors were shut where the disciples were assembled for fear of the Jews, came Jesus and stood in the midst, and saith unto them, Peace be unto you. And when he had so said, **he shewed unto them his hands and his side. Then were the disciples glad, when they saw the Lord.**

Acts 1:9 And when he had spoken these things, while they beheld, **he was taken up; and a cloud received him out of their sight**.

Philippians 3:20, 21 For our conversation is in heaven; from whence also we look for the Saviour, the Lord Jesus Christ: Who shall change our vile body, that it may be **fashioned like unto his glorious body**, according to the working **whereby he is able even to subdue all things unto himself.**

Since knowledge, love, and doing good at our jobs seems to be essential to reaching that destiny, the Sons of God must have functions similar to those of humans on Earth. We have some clues. The spirit that we call Jesus Christ is the light of the world. Jesus while on Earth prayed to the Father. Satan is god of this world, the Earth. Putting these clues together brought me to the conclusion that since the Sun is the light of the world—the Sun, the spirit of Jesus Christ must be responsible for the entire Solar System. This spirit controls the activities of, in, and on the Sun and all the planets and satellites. Satan is the spirit that, also doing the will of God, controls some activities of the planet Earth. Jesus is superior to Satan and can overrule him, if that ever is necessary. The term "God" includes all the spirits that exist in the entire Universe; but to us, that means merely that God includes what Christians call *The Holy Trinity.* Therefore, the destiny of individual souls that continue developing until the end of their spiritual growth will be assigned responsibility for celestial bodies. In between what we are now and when we reach the level of Sons of God, we will be gradually given responsibilities for other jobs on Earth, somewhere else in the Solar System, or somewhere else in the Universe. Perhaps along the way to being Sons of God, we will perform the duties of guardian angels.

Many Mansions

Jesus said, "*In my father's house are many mansions…*" (John 14:1-3) Bible scholars usually say this means some kind of spiritual existence. Some people believe it refers to many bodies that a soul occupies during reincarnations. Jesus did not elaborate on this subject; however, a logical explanation could be that there are other planets where there are intelligent beings that are part of the Father's creation. Saying there are many "mansions" implies that there are residents in them. Legends, ancient carvings and pictures, and anecdotal reports from many individuals tell us that some of these residents might have actually visited our planet during ancient times and into the 21st century. For reasons unknown to us, these aliens, if there really were extraterrestrial visitors, have not made themselves known to the world in general. (They might be afraid of us, just as we fear vicious animals.) If intelligent beings live on other planets within our galaxy, and in other galaxies, they probably are destined for the same position in the spiritual world that we are.

When enough of the intelligent beings in the Universe reach the appropriate level of spiritual energy, the Universe will be returned to its center, either to be recycled into a new Universe or to generate one or more, parallel Universes.

And, with that, I will leave you to your own search for the destiny of all mankind. Love the Lord thy God with all thy heart, and with all thy soul, and with all thy mind, and with all thy strength; and love thy neighbor as thyself. Continue to acquire knowledge. And accept the peace of God.

> **Philippians 4:7-9** And the peace of God, which passeth all understanding, shall keep your hearts and minds through Christ Jesus. Finally, brethren, whatsoever things are true, whatsoever things are honest, whatsoever things are just, whatsoever things are pure, whatsoever things are lovely, whatsoever things are of good report; if there be any virtue, and if there be any praise, think on these things. Those things, which ye have both learned, and received, and heard, and seen in me, do: and the God of peace shall be with you.

APPENDIX A
REINCARNATION IN THE BIBLE

The Internet has listings of many verses in the Bible that seem to discuss reincarnation. One source listed 40 Biblical references that might refer to reincarnation. Some sources criticize the concept and claim the references listed by advocates of reincarnation are the result of selecting isolated passages and deliberately misinterpreting them. I admit some of the references can be interpreted in several ways and the advocates probably were over enthusiastic to prove their cause. Nevertheless, some passages in the Bible can reasonably be interpreted to support the concept.

Some promises in the Bible are meaningless in the traditional understanding without some method of fulfilling the promises. One example is the third Beatitude in Jesus Christ's Sermon on the Mount. The meek probably could not inherit the earth in the lives when they are meek; therefore, the inheritance most logically occurs in future lives in earthly existences. Traditional Christians say the inheritance occurs in Heaven; however, Heavenly rewards will not be on the Earth, but in a spiritual existence, therefore they could not inherit the earth. A related verse in Isaiah states the poor among men will rejoice in God. This is usually interpreted to mean a spiritual reward in Heaven; but it could also mean the rejoicing occurs in future incarnations.

> **Matthew 5:5** Blessed (honored by God) are the meek: for **they shall inherit the earth**.

> **Isaiah 29:19** The meek also shall increase their joy in the LORD, and the poor among men shall rejoice in the Holy One of Israel.

Luke 6:38 is another example of future rewards for good deeds done during a lifetime. The verse promises that what a person does will be rewarded in a manner similar to what was given. In many cases, especially the case of missionaries who spent entire lifetimes helping people frequently end their lives close to poverty. The verse does not say the good will be rewarded in heaven—it states that the reward will be in "*the same measure*" that was "meted" to others. This indicates that the reward will be on Earth.

> **Luke 6:38** Give, and it shall be given unto you; good measure, pressed down, and shaken together, and running over, shall men give into your bosom. For with **the same measure that ye mete withal it shall be measured to you** again.

The following verse states that evil deeds will be punished in the same manner that they used to commit their evil deeds. A person that kills with a sword could not be killed with a sword in Heaven, since I doubt that there are any physical swords there. Most tyrants who kill thousands and millions of people will not perish in the same way they harmed others in the lifetime that they did wrong. Therefore, the recompense will occur in some future lifetime. I assume those who used a sword in previous times will be killed with modern ways that have the same effect on a person's life. The killing would have to be done in a future lifetime. The last sentence in the verse states that the patience of saints who have perished by the swords of the evildoers will be satisfied when they see their former enemies killed in the same manner. This seems to contradict the admonition to love our enemies; but this sentence means that our patience in loving them in one lifetime and failing to take revenge will be proven correct, proving that God knows best how to correct those who do wrong. The effect on the evildoers of being killed is what is important. They must be able to understand they are suffering in one life because of what they have done in the past.

> **Revelation 13:10** He that leadeth into captivity shall go into captivity: he that killeth with the sword must be killed with the sword. Here is the patience and the faith of the saints.

Only those who are ready for the knowledge can understand the teachings of Jesus. He explained one parable to the disciples in the following passage. There is no way of discovering what other secrets Jesus trusted them with. I believe the secret of reincarnation was one of those that Jesus explained to his disciples in private and left the rest of those who heard him to learn for themselves what he really meant. If people are ready to know, they will know what he meant. I think that Gnosticism was based on this concept, just as is the concept of predestination in modern Christian Churches—only those who have developed their souls to a level high enough so they can actually understand what the Bible means; and only they will be "saved."

> **Matthew 13:10-17** And the disciples came, and said unto him, Why speakest thou unto them in parables? He answered and said unto them, Because it is given unto you to know the mysteries of the kingdom of heaven, but to them it is not given. For whosoever hath, to him shall be given, and he shall have more abundance: but whosoever hath not, from him shall be taken away even that he hath. Therefore speak I to them in parables: because they seeing see not; and hearing they hear not, neither do they understand. And in them is fulfilled the prophecy of Esaias, which saith, By hearing ye shall hear, and shall not understand; and seeing ye shall see, and shall not perceive:
>
> For this people's heart is waxed gross, and their ears are dull of hearing, and their eyes they have closed; lest at any time they should see with their eyes, and hear with their ears, and should understand with their heart, and should be converted, and I should heal them. But blessed are your eyes, for they see: and your ears, for they hear. For verily I say unto you, That many prophets and righteous men have desired to see those things which ye see, and have not seen them; and to hear those things which ye hear, and have not heard them.

Jesus criticized the high priests and the elders of the people because they failed to accept his lessons, while the "*publicans and the harlots*" who had accepted him were farther along the path toward perfection and closer to the kingdom of God than the so-called "holy men" were.

> **Matthew 21:31, 32** Whether of them twain did the will of his father? They say unto him, The first. Jesus saith unto them, Verily I say unto you, That the publicans and the harlots go into the kingdom of God before you. For John came unto you in the way of righteousness, and ye believed him not: but the publicans and the harlots believed him: and ye, when ye had seen it, repented not afterward, that ye might believe him.

In the following parable, Jesus indicates that talents (inborn skills) would be given to every man "*according to his several ability.*"

The servants were supposed to enlarge their talents; this was an opportunity, as well as a test. It seems that the amount of opportunities individuals receive are given based on how far along they are to the kingdom of God.

> **Matthew 25:14, 15** For the kingdom of heaven is as a man travelling into a far country, who called his own servants, and delivered unto them his goods. And unto one he gave five talents, to another two, and to another one; to every man **according to his several ability**; and straightway took his journey.

If being born in luxury is a reward for having lived a previous life in the proper way, why did Jesus say it was difficult for a rich man to enter into the kingdom of God?

> **Mark 10:23-27** And Jesus looked round about, and saith unto his disciples, **How hardly shall they that have riches enter into the kingdom of God!** And the disciples were astonished at his words. But Jesus answereth again, and saith unto them, Children, how hard is it for them that trust in riches to enter into the kingdom of God! It is easier for a camel to go through the eye of a needle, than for a rich man to enter into the kingdom of God. And they were astonished out of measure, saying among themselves, Who then can be saved? And Jesus looking upon them saith, With men it is impossible, but not with God: for with God all things are possible.

Jesus told Peter that he would be rewarded for his sacrifices in "*this* time." Peter was promised a reward in some future time before being in the "*world to come.*" "This time" means the time before the last days of this civilization. Peter was promised rewards in a future incarnation—reincarnation on Earth in this time cycle.

> **Mark 10:28-31** Then Peter began to say unto him, Lo, we have left all, and have followed thee. And Jesus answered and said, Verily I say unto you, There is no man that hath left house, or brethren, or sisters, or father, or mother, or wife, or children, or lands, for my sake, and the gospel's, **But he shall receive an hundredfold now in this time**, houses, and brethren, and sisters, and mothers, and children, and lands, with persecutions; and **in the world to come eternal life**. But many that are first shall be last; and the last first.

"*Every eye*" includes those who were alive at the time of Jesus' death. The people who were responsible for the crucifixion and death of Jesus of Nazareth will be alive in the flesh, reincarnated, when Jesus returns and the end of our present civilization comes.

> **Revelation 1:7** Behold, he cometh with clouds; and **every eye** shall see him, **and they also which pierced him:** and all kindreds of the earth shall wail because of him. Even so, Amen.

The souls that keep the words of Jesus will not be incarnated during the "*hour of temptation.*"

> **Revelation 3:10-13** Because thou hast kept the word of my patience, I also will keep thee from the **hour of temptation**, which shall come upon all the world, to try them that dwell upon the earth. Behold, I come quickly: hold that fast which thou hast, that no man take thy crown. Him that overcometh will I make a pillar in the temple of my God, and he shall go no more out: and I will write upon him the name of my God, and the name of the city of my God, which is new Jerusalem, which cometh down out of heaven from my God: and I will write upon him my new name. He that hath an ear, let him hear what the Spirit saith unto the churches.

As mentioned earlier in the main text, the entire book of Job seems to be discussing reincarnation; however, the following passage seems a distinct reference to reincarnation. Job is praying that he not be alive during the time of God's wrath. Verses 24 and 25 specifically mention "*living again.*" The following words refer to Job waiting for his "*appointed time*" when his new body (his change) will be given to him. God will expect the reincarnated Job to work for Him.

> Job 14:13-15 O that thou wouldest **hide me in the grave, that thou wouldest keep me secret, until thy wrath be past**, that thou wouldest appoint me a set time, and remember me! **If a man die, shall he live again**? all the days of my appointed time will **I wait, till my change come**. Thou shalt call, and **I will answer thee: thou wilt have a desire to the work of thine hands**.

Solomon seems to have discussed the results of reincarnation, especially where he wrote the following passage. Ecclesiastes 1:3-11 might be describing the concept of humans reincarnating. Solomon wrote about the repetitive cycles of nature—reincarnations of natural events; and in the last verses he described the condition of being in a new body in a new era, which is typical of people who attempt to "remember" previous incarnations. Some children mention events or places that they have never seen or heard about and some of them display talents that normally require years of study and practice; but as they get older, many of them lose those thoughts, or memories. Some child prodigies lose their remarkable abilities as they get older. New incarnations are meant to give us new opportunities for advancement of our souls; that is why we are not supposed to remember previous lives.

> **Ecclesiastes 1:3-11** What profit hath a man of all his labour which he taketh under the sun? One generation passeth away, and another generation cometh: but the earth abideth for ever. The sun also ariseth, and the sun goeth down, and hasteth to his place where he arose. The wind goeth toward the south, and turneth about unto the north; it whirleth about continually, and the wind returneth again according to his circuits. All the rivers run into the sea; yet the sea is not full; unto the place from whence the rivers come, thither they return again. All things are full of labour; man cannot utter it: the eye is not satisfied with seeing, nor the ear filled with hearing. The thing that hath been, it is that which shall be; and that which is done is that which shall be done: and there is no new thing under the sun.
>
> Is there any thing whereof it may be said, See, this is new? it hath been already of old time, which was before us. **There is no remembrance of former things; neither shall there be any remembrance of things that are to come with those that shall come after**.

BIBLIOGRAPHY

The books listed in this bibliography are the most important sources of my information and most of them are recommended for anyone that wants to know more about our destiny. I am sure a lot of my book might also be traced to other places over the years: sermons: including some delivered by my son, Jeffrey, newspaper or magazine articles, private discussions with people that knew more than I did, and many other sources that I do not remember. I don't think I quoted anyone directly—at least, I hope I didn't. Please do not assume that any of the authors of any of these sources might agree with me—I am sure many of them would find my conclusions not entirely to their liking.

Human Destiny. By Pierre Lecomte du Nouy. Longmans, Green and Co. New York, London, and Toronto. 1947.

The Purpose Driven Life. By Rick Warren. Zondervan, Grand Rapids, Michigan. 2002.

The Case for a Creator. By Lee Strobel. Zondervan, Grand Rapids, Michigan. 2004.

Nelson's New Illustrated Bible Dictionary. General Editor, Robert E, Youngblood. Thomas Nelson Publishers; Nashville, Tennessee. 1995.

Halley's Bible Handbook. By Henry H. Halley. Zondervan Publishing House; Grand Rapids, Michigan. 1962.

The Bible as History. By Werner Keller. Barnes and Noble Books. 1980.

Interpreting the Bible. By A. Berkeley Mickelsen. Wm. B. Eerdmans Publishing Co. Grand Rapids, Michigan. 1976.

The Continuity of the Bible—the Patriarchs. By Thomas Linton Leishman. The Christian Science Publishing Society, Boston. 1968.

The World's Religions. Edited by Sir Norman Anderson. William B. Eerdmans Publishing Co. Grand Rapids, Michigan. 1991.

Larouse Encyclopedia of Mythology. Prometheus Press, New York. 1960.

God, a Brief History. By John Bowker. D. R. Publishing, Inc. London, New York. 2002.

Exploring Religion. By Roger Schmidt. Wadsworth, Inc. Belmont, California. 1980.

World Religions. Edited by Martin Palmer. Barnes and Noble Books, New York. 2002.

The World's Great Religions. Edited by Sam Welles. Time Inc. New York. 1957.

The Wisdom of China and India. Edited by Lin Yutang. The Modern Library, New York. 1942.

Our Oriental Heritage. By Will Durant. Simon and Schuster, New York. 1954.

The Art of Loving. By Erich Fromm. Harper Colophon Books, N. Y. and Evanston. 1956.

Intimate Behavior. By Desmond Morris. Random House, New York. 1971.

Bhagavada-Gita As It Is. By A.C. Bhaktivedanta Swami Prabhaupada. The Bhaktivedanta Book Trust, Los Angeles, Ca. 1972.

Our Evolving Universe. By Walter H. Hesse. Dickenson Publishing Co. Encino, Ca. 1997.

Relativity for the Million, By Martin Gardner. Macmillan edition, 1962. Pocket Books, Pocket Cardinal edition, New York, N. Y. 1965.

The Ancient Engineers. By Sprague De Camp. Barnes and Noble Books, New York. 1993.

Plato-Five Great Dialogues. Translated by B. Jowett. Walter J. Black; New York. 1942.

Atlantis, the Antediluvian World. Edited by Egerton Sykes. Grammercy Publishing, N. Y. 1949.

Frames of Mind—The Theory of Multiple Intelligences. By Howard Gardner. Basic Books. 1993.

The Bell Curve, Intelligence and Class Structure in American Life. By Richard Hernstein & Charles Murray. The Free Press. 1994.

Destiny of Souls—New Case Studies of Life between Lives. By Michael Newton Ph.D. Llewellyn Publications, St. Paul, Minnesota. Sixth printing, 2002. First publication 1931.

Reincarnation—The Missing Link in Christianity. By Elizabeth Clare Prophet, with Erin L. Prophet. Summit University Press. 1997.

Reincarnation and Immortality. By Rudolf Steiner. Harper & Row Publishers, San Francisco. 1970.

Life after Life. By Raymond A. Moody Jr. Bantam Books, New York, Toronto, and London. 1978.

Reflections on Life after Life. By Raymond A. Moody Jr. Bantam Books, New York, Toronto, and London. 1978.

90 Minutes in Heaven. By Don Piper with Cecil Murphy. Revell, Grand Rapids, Michigan. 2004.

Human Personality and its Survival of Bodily Death. By F. W. H. Myers. Edited by Susy Smith. University Books, Inc. Hyde Park, N. Y. 1961.

The Prophecies of Nostradamus, Translated, Edited, and Introduced by Erika Cheetham. A Wideview/Perigee Book. 1973.

Come be My Light: The Private Writings of the Saint of Calcutta, edited by Brian Kolodiejchuk, Doubleday. 2007.

Unlocking the Secrets of the Shroud. By Gilbert R. Lavoie, M. D. Thomas More, Allen Texas. 1998.

DESTINY, Volume One, Our Father's Commandments, A Comparative Analysis of the Commandments and the Lord's Prayer. By Edward Nagell. Infinity Publishing.com, West Conshohocken, PA. 2006.

DESTINY, Volume Two, God, Creation, and Religion, God's Actions and Mankind's Reactions. By Edward Nagell. Infinity Publishing.com, West Conshohocken, PA. 2006.

DESTINY, Volume Three, Christianity. The Do-It-Yourself Religion, By Edward Nagell. Infinity Publishing.com, West Conshohocken, PA. 2006.

DESTINY, Volume Four, Human Societies. Prosperity, Pride, and Poverty, the Endless Cycle. By Edward Nagell. Infinity Publishing.com, West Conshohocken, PA. 2006.

I realize most of what I have presented in this book may be a new concept of the Universe and the relationship of Humans to the Universe. If these ideas are new to you, please do not resist the concepts without analyzing them in light of your previous learning. The final thought for this discussion comes from John Locke, an English philosopher. (1632-1704)

> "New opinions are always suspected, and usually opposed, without any other reason but because they are not already common."

I am grateful that you have read through this book and hopefully through the first four volumes of the Destiny series. I sincerely pray that you have profited from reading my thoughts. Keep studying and keep praying for continuous spiritual guidance and growth.

Edward Nagell 2008

Made in the USA
Monee, IL
08 November 2026

67971934R00138